INNOVATION IN DESIGN, COMMUNICATION AND ENGINEERING

PROCEEDINGS OF THE 3RD INTERNATIONAL CONFERENCE ON INNOVATION, COMMUNICATION AND ENGINEERING (ICICE 2014), GUIYANG, GUIZHOU, P.R. CHINA, 17–22 OCTOBER 2014

# Innovation in Design, Communication and Engineering

*Editors*

## Teen-Hang Meen
*National Formosa University, Taiwan*

## Stephen D. Prior
*The University of Southampton, UK*

## Artde Donald Kin-Tak Lam
*Fuzhou University, China*

**CRC Press**
Taylor & Francis Group
Boca Raton London New York Leiden

CRC Press is an imprint of the
Taylor & Francis Group, an **informa** business

A BALKEMA BOOK

*CRC Press/Balkema is an imprint of the Taylor & Francis Group, an informa business*

© 2015 Taylor & Francis Group, London, UK

Typeset by V Publishing Solutions Pvt Ltd., Chennai, India

Published by: CRC Press/Balkema
P.O. Box 11320, 2301 EH Leiden, The Netherlands
e-mail: Pub.NL@taylorandfrancis.com
www.crcpress.com – www.taylorandfrancis.com

ISBN: 978-1-138-02752-7 (Hbk + CD-Rom)
ISBN: 978-1-315-68761-2 (eBook PDF)

*Innovation in Design, Communication and Engineering – Meen, Prior & Lam (Eds)*
*© 2015 Taylor & Francis Group, London, ISBN 978-1-138-02752-7*

# Table of contents

*Design theory & knowledge innovation*

# Editorial

This volume represents the proceedings of the 2014 International Conference on Innovation, Communication and Engineering (ICICE 2014). This conference was organized by the Taiwanese Institute of Knowledge Innovation, and held October 17–22, 2014 in Guiyang, Guizhou, P.R. China. The conference received 523 submitted papers from 8 countries, whereby 170 papers were selected by the committees to be presented at ICICE 2014. The conference provided a unified communication platform for researchers in a wide range of disciplines, from information technology, communication science and applied mathematics to computer science, advanced material science and engineering. This proceedings volume enables inter-disciplinary collaboration between science and engineering technologists in the academic and industry fields as well as networking internationally.

EDITORS:

*Teen-Hang Meen*

Dr. Teen-Hang Meen
National Formosa University, Taiwan

*S Prior*

Dr. Stephen D. Prior
The University of Southampton, UK

Dr. Artde Donald Kin-Tak Lam
Fuzhou University, China

*Advanced material science & engineering*

*Innovation in Design, Communication and Engineering – Meen, Prior & Lam (Eds)*
© 2015 Taylor & Francis Group, London, ISBN 978-1-138-02752-7

# The characteristic of Ga-doped ZnO nanorods photodetectors

S.J. Young, C.L. Chiou & T.H. Meen
*Department of Electronic Engineering, National Formosa University, Yunlin, Taiwan*

Y.H. Liu
*Department of Electrical Engineering, Institute of Microelectronics, National Cheng Kung University, Tainan, Taiwan*

L.W. Ji
*Institute of Electro-Optical and Materials Science, National Formosa University, Yunlin, Taiwan*

ABSTRACT:    In this study, high-density single crystalline Ga-doped ZnO (GZO) nanorods were grown on glass substrate by the hydrothermal method. The structural and optoelectronic properties of Ga-doped ZnO nanorods are studied. The microstructure of the GZO was studied by Scanning Electrical Microscope (SEM). The structural characteristics of the GZO were measured by X-Ray Diffraction (XRD). It was found that the peaks related to the wurtzite structure ZnO (100), (002), and (101) diffraction peaks. The (002) peak indicates that the nanorods were preferentially oriented in the c-axis direction. The existence of Ga was examined by Energy Diffraction Spectra (EDS), indicating Ga atom entered into the ZnO lattice. The optical properties of the GZO were measured by photoluminescence spectra. It was found that the Ultraviolet (UV) emission located near 380 nm is dominant in the PL spectrum, which is attributed to the near band gap excitonic emission. A GZO nanorods Metal-Semiconductor-Metal (MSM) ultraviolet Photodetector (PD) was also fabricated. The photocurrent to dark-current constant ratio of the fabricated PD was approximately 15.2 when biased at 1 V.

# Deposition of $In_2O_3$-$Al_2O_3$-$SnO_2$ (IATO) transparent conduction thin films using non-vacuum method

Chien-Chen Diao
*Department of Electronic Engineering, Kao Yuan University, Kaohsiung, Taiwan, R.O.C.*

Chia-I Chuang, Siou-Mei Huang & Cheng-Fu Yang
*Department of Chemical and Materials Engineering, National University of Kaohsiung, Taiwan, R.O.C.*

Sung-Mao Wu
*Department of Electrical Engineering, National University of Kaohsiung, Taiwan, R.O.C.*

Yuan-Tai Hsieh
*Department of Electronic Engineering, Southern Taiwan University, Tainan, Taiwan, R.O.C.*

ABSTRACT:  The $In_2O_3$-$Al_2O_3$-$SnO_2$ (In:Al:Sn = 68.8:11.8:8.4, IATO) transparent conduction thin films were synthesized by using the non-vacuum Spray Coating Method (SCM). At first, the IATO power was mixed, calcined at 1100°C, and ground into nano-scale particles, then the 6 wt% IATO particles were dispersed into isopropyl alcohol (IPA) to get the solution for SCM. 0.1 ml IATO solution was sprayed on the 2 cm × 1 cm glass substrates and then the nano-scale IATO solution was carried out the thermal treatment under different temperatures in a furnace in air. The annealing temperature was changed from 200°C to 800°C and the annealing time was 3 h. After annealing, the influences of the annealing tempera-ture on crystallization of the IATO thin films were investigated. The surface and cross-section morpholo-gies, optical transmission ratio, resistivity ($\rho$), hall mobility ($\mu$), and carrier concentration of the IATO thin films were well investigated in this study.

*Keywords*:  $In_2O_3$-$Al_2O_3$-$SnO_2$; transparent conduction thin film; spray coating method

*Innovation in Design, Communication and Engineering – Meen, Prior & Lam (Eds)*
*© 2015 Taylor & Francis Group, London, ISBN 978-1-138-02752-7*

# Deposition of ZnO-In$_2$O$_3$-Ga$_2$O$_3$ (IGZO) thin films by using Spray Coating Method

Chin-Guo Kuo
*Department of Industrial Education, National Taiwan Normal University, Taipei, Taiwan, R.O.C.*

I-Chun Lin, Chia-I Chuang, Cheng-Fu Yang & Hsuan-Yu Yang
*Department of Chemical and Materials Engineering, National University of Kaohsiung, Taiwan, R.O.C.*

Chia-Ching Wu
*Department of Electronic Engineering, Kao Yuan University, Kaohsiung, Taiwan, R.O.C.*

ABSTRACT:   In this study, the micro-scale of In$_2$O$_3$, Ga$_2$O$_3$, and ZnO oxide powders (with mole ratio In$_2$O$_3$: Ga$_2$O$_3$: ZnO = 1:1:2, abbreviated as IGZO) were put into a nano-ground machine with the addition of 1 wt% KD1 dispersant for nano-scale IGZO particles' dispersion. Then the 6 wt% IGZO particles were dispersed into Isopropyl Alcohol (IPA) to get the solution for SPM to prepare the IGZO thin films. 0.1 ml IGZO solution was sprayed on the 2 cm × 1 cm glass substrates and then the nano-scale IGZO solution was carried out for thermal treatment under different temperatures in a furnace in air. The annealing temperature was changed from 200°C to 700°C and the annealing time was 3 h. The effects of the annealing temperature on the surface and cross-session morphologies, X-ray diffraction pattern, optical transmission spectrum, carrier concentration, carrier mobility, and resistivity of the IGZO thin films on glass substrates were studied. The measured results showed that the annealing temperature had large effect on the characteristics of the IGZO thin films.

*Keywords*:   IGZO; nano-scale; Spray Coating Method; annealing

*Innovation in Design, Communication and Engineering – Meen, Prior & Lam (Eds)*
© *2015 Taylor & Francis Group, London, ISBN 978-1-138-02752-7*

# Growth of $Fe_2O_3$ thin films and $Fe_2O_3$-doped ZnO micron-flower by using modified Spray Coating Method

Cheng-Yi Chen
*Department of Electrical Engineering, Cheng Shiu University, Kaohsiung, Taiwan, R.O.C.*

Siou-Mei Huang, I-Chun Lin & Cheng-Fu Yang
*Department of Chemical and Materials Engineering, National University of Kaohsiung, Taiwan, R.O.C.*

Sung-Mao Wu
*Department of Electrical Engineering, National University of Kaohsiung, Taiwan, R.O.C.*

Yuan-Tai Hsieh
*Department of Electronic Engineering, Southern Taiwan University, Tainan, Taiwan, R.O.C.*

ABSTRACT: $Fe_2O_3$-doped ZnO thin films can be used as the multiferroic materials, which show simultaneous ferroelectric and magnetic ordering, exhibit unusual physical properties, and in turn, promise new device applications. In this study, iron nitrate [$Fe(NO_3)_3 \cdot 6H_2O$] and zinc nitrate [$Zn(NO_3)_2 \cdot 6H_2O$] were mixed with deionized (D. I.) water to form the 1 mol% $Fe_2O_3$ and 1 mol% Fe-doped ZnO solution (ZnO: $Fe_2O_3 = 95:5$). $Fe_2O_3$ thin films and Fe-doped ZnO thin films were hoped to be prepared by using the modified Spray Coating Method (m-SCM). $Fe_2O_3$ solution was annealed from 300°C to 700°C and Fe-doped ZnO solution was annealed from 200°C to 700°C for densification and crystallization. We found that $Fe_2O_3$ thin films formed a cracked-piece structure and Fe-doped ZnO formed a micron-flower structure. The influences of the annealing temperature on surface morphology, resistivity ($\rho$), hall mobility ($\mu$), and carrier concentration of $Fe_2O_3$ thin films were investigated. The influence of the annealing temperature on surface morphology of Fe-doped ZnO micron-flowers was also investigated.

*Keywords*: ZnO; micron-flower; Spray Coating Method; thin film

# Green molding compounds for encapsulating microelectronic devices

Chean-Cheng Su, Cheng-Fu Yang & Ping-Lin Wu
*Department of Chemical and Materials Engineering, National University of Kaohsiung, Kaohsiung, Taiwan, R.O.C.*

Song-Mao Wu
*Department of Electrical Engineering, National University of Kaohsiung, Kaohsiung, Taiwan, R.O.C.*

ABSTRACT: The Epoxy Molding Compound (EMC) with biphenyl resins and highly loaded fillers can retard flammability and is a green material. A highly loaded filler is critical to producing reliable packaging materials for microelectronic devices, with high flame retardation, high thermal resistance, high moisture resistance, favorable mechanical properties and a low thermal expansion coefficient of EMC. In this work, a new organophosphine thermally latent accelerator, triphenyl phosphine-benzoquino (TPP-BQ), was synthesized with variously sized TPP-BQ particles. It was used in the high filler-loaded EMCs that were based on the biphenyl epoxy. EMCs with 92 wt% filler contents are accelerated by the latent heat accelerator, TPP-BQ, which has an excellent flowability in the molding process. Additionally, the effects of the latent accelerator on the activity, thermal behavior, rheology and molding characteristics of EMCs were investigated.

*Innovation in Design, Communication and Engineering – Meen, Prior & Lam (Eds)*
© 2015 Taylor & Francis Group, London, ISBN 978-1-138-02752-7

# Investigation of a plastic injected container of PCM for temperature rapid–balancing and long–maintaining

Chang-Ren Chen & Tsung-Nan Wu
*Department of Mechanical Engineering, Kun Shan University, Tainan, Taiwan*

ABSTRACT: In this study, a Phase Change Material (PCM) container made of plastic by injection molding process has been designed in a suitable structure to store PCMs in combination with a vacuum layer acting as an excellent insulated outer cover. The target of this design is to create a rapid temperature balancing container which can firstly absorb heat from the stored substance to change its temperature within several minutes into a desired value, set by the melting temperature and type of used PCMs, and then keep that temperature last longer as the absorbed heat is later released. Models in different shape using a commercial PCM (called PW-63) with melting temperature of 63°C have been made and tested carefully in laboratory. The best results, derived from the multi-layer structure, have shown that average temperature of boiled water (stored substance) reached the desired range of 55~65°C within only about 15 minutes and this temperature range lasted within nearly 5 hours. This time is even much longer for lower temperature desired value. This is a big advantage compared with vacuum container without PCM (in which the desired range was reached after about 4 hours but lasted for only more than 2 hours), which is normally considered a very good device for isolating storage. With this good performance, the plastic multi-layer PCM container can become a good solution for stainless steel PCM container in heat transfer enhancement in latent heat storage applications using PCMs. It also brings along other advantages compared with stainless steel container such as lighter weight, cheaper, faster and easier mass-production by means of plastic injection process.

# Study on using commercial touch panel film as the gate electrode for ISFET-based pH sensor

Yung-Chen Wu

*Department of Mechanical and Electro-Mechanical Engineering, National Sun Yat-Sen University, Kaohsiung, Taiwan*
*Metal Industries Research and Development Centre, Kaohsiung, Taiwan*

Shang-Jing Wu & Che-Hsin Lin

*Department of Mechanical and Electro-Mechanical Engineering, National Sun Yat-Sen University, Kaohsiung, Taiwan*

ABSTRACT: This study developed a high performance Ion-Sensitive Field-Effect-Transistor (ISFET) based pH sensor utilizing commercial Touch Panel Film (TPF) as the gate electrode. The TPF was composed of a PET substrate both sides symmetrically coated with a niobium pentoxide, a silicon dioxide and an indium tin oxide (ITO/SiO$_2$/Nb$_2$O$_5$) layers by industrial roll-to-roll sputtering process, which qualified by industrial quality specifications was approved suitable for commercialized products development. The gate electrode was connected to a MOSFET by a plug-in design such that the developed pH sensor could be replaced in seconds. From the measurements, the TPF-based pH sensor had high sensitivity of 59.2 mV/pH with R$^2$ value of 0.9948 in buffer solution pH range between 3 and 13, which was almost identical to the Nernst theoretical response. Meanwhile the response of the developed sensor was fast (1s) with good stability (C.V. <1%). This study had developed a flexible TPF-based pH sensor which was of high performance and suitable for disposable biosensor applications.

*Innovation in Design, Communication and Engineering – Meen, Prior & Lam (Eds)*
*© 2015 Taylor & Francis Group, London, ISBN 978-1-138-02752-7*

# Thermoelectric properties of the difference between $Zn_{3.9}Sb_3$ and $Zn_4Sb_3$

B.J. Tsai & K.C. Cheng
*Department of Mechanical Engineering, Chung Hua University, Hsinchu, Taiwan*

ABSTRACT: The ratio between Zn and Sb has been studied and optimized in this thermoelectricity study. The ratio was changed from 4:3 to 3.9:3 based on the different melting points of Zn (419.5°C) and Sb (630.8°C). Thermoelectric properties including Seebeck value(s), conductivity(s) and power factor have been investigated by a different hot pressing temperature and time. When $Zn_{3.9}Sb_3$ alloy increased melting temperature and time will increase density and conductivity from power metallurgy. In this investigation, the difference of the melting temperature between Zn and Sb, causing a different volatilizing speed, and improving thermoelectric effect by $Zn_{3.9}Sb_3$ structure is discussed.

*Innovation in Design, Communication and Engineering – Meen, Prior & Lam (Eds)*
*© 2015 Taylor & Francis Group, London, ISBN 978-1-138-02752-7*

# Investigation of effective parameters on mechanical property in nanoindentation of polymer/carbon nanotubes nanocomposite using square representative volume element

Minh-Tai Le & Shyh-Chour Huang

*Department of Mechanical Engineering, National Kaohsiung University of Applied Sciences, Kaohsiung, Taiwan, R.O.C.*

ABSTRACT:   Characterizing mechanical properties of Polymer/Carbon nanotubes nanocomposites has always been a challenge. The present research is to develop a representative volume element method for modeling of nanoindentation behavior of Epoxy/Single Wall Carbon Nanotubes (SWCNTs) nanocomposite that has nonlinear elastic behavior using numerical analysis. A series of two-dimensional nanoindentation simulations were carried out to calculate and evaluate its elastic modulus and hardness. The effect of volume percentages of the SWCNTs on mechanical property was investigated. Besides, the influences of indenter geometry as well as friction coefficient on the response of nanoindentation were also mentioned.

*Innovation in Design, Communication and Engineering – Meen, Prior & Lam (Eds)*
© 2015 Taylor & Francis Group, London, ISBN 978-1-138-02752-7

# The performance of polymer solar cells by various fabrication parameters and ITO

Jhong-Ciao Ke & Yeong-Her Wang
*Department of Electrical Engineering, Institute of Microelectronics,
National Cheng-Kung University, Tainan, Taiwan*

Kan-Lin Chen
*Department of Electronic Engineering, Fortune Institute of Technology, Kaohsiung, Taiwan*

Chien-Jung Huang
*Department of Applied Physics, National University of Kaohsiung, Nanzih, Kaohsiung, Taiwan*

Neng-Lang Shih
*Department of Life Sciences, National University of Kaohsiung, Nanzih, Kaohsiung, Taiwan*

Chern-Hwa Chen
*Department of Civil and Environmental Engineering, National University of Kaohsiung, Nanzih,
Kaohsiung, Taiwan*

ABSTRACT: The performance of polymer solar cells based on poly[N-900-hepta-decanyl-2,7-carbazole-alt-5,5-(40,70-di-2-thienyl-20,10,30-benzothiadiazole)] (PCDTBT) and [6,6]-phenyl C71-butyric acid methyl ester ($PC_{71}BM$) made at the rotation speed of 1000~3000 rpm was studied. The structure of device is Indium Tin Oxide (ITO)/molybdenum oxide (10 nm)/PCDTBT: $PC_{71}BM$ (1000~3000 rpm)/2,9-dimethyl-4,7-diphenyl-1,10-phenanthroline (10 nm)/aluminum. The maximum power conversion efficiency of the device is obtained at a rotation speed of 2000 rpm, which is attributed to reach the optimal balance between the light absorption and the carrier collection efficiency. In addition, the different sheet resistances of ITO were used as substrate to find out which is appropriate for the organic solar cell application. The result showed that the device using 15 Ω/sq of ITO as substrate has a better efficiency than that of the device using 7 Ω/sq of ITO. It is due to the transmittance of ITO (15 Ω/sq) being higher than that of ITO (7 Ω/sq), especially from 350 nm to 475 nm, that leads to the efficiency of the device (2000 rpm) improving from 3.84% to 4.25%.

*Innovation in Design, Communication and Engineering – Meen, Prior & Lam (Eds)*
*© 2015 Taylor & Francis Group, London, ISBN 978-1-138-02752-7*

# The impact on ITO by wet etching for flexible organic solar cells

Pao-Hsun Huang
*Department of Applied Physics, National University of Kaohsiung, Nanzih, Kaohsiung, Taiwan*

Kan-Lin Chen
*Department of Electronic Engineering, Fortune Institute of Technology, Kaohsiung, Taiwan*

Chien-Jung Huang
*Department of Applied Physics, National University of Kaohsiung, Nanzih, Kaohsiung, Taiwan*

Teen-Hang Meen
*Department of Electronic Engineering, National Formosa University, Taiwan*

ABSTRACT: The result of patterned ITO by wet etching was researched and used to fabricate on the flexible substrate (PET) for Small Molecule Organic Solar Cells (SM-OSCs) at the same time. The device was based on Copper Phthalocyanine (CuPc) as donor and fullerene ($C_{60}$) as acceptor with a basic structure of PET-ITO//CuPc (20 nm)/$C_{60}$(30 nm)/BCP (10 nm)/Aluminum (Al). It is found that surface roughness of a non-conducting area has a wide difference after various etching times but the short-circuited measurement is the same. The optimal root mean square roughness ($R_{rms}$) of the non-conducting area was obtained and also affected by both appropriate etchant ($H_2C_2O_4$ liquor) and crystallization morphology of ITO. The power conversion efficiency of the device was improved from 0.1% to 0.4% due to the decrease of $R_{rms}$ from 89.9 nm to 2.2 nm.

*Keywords:* organic solar cell; wet etching; ITO; flexible substrate

*Innovation in Design, Communication and Engineering – Meen, Prior & Lam (Eds)*
*© 2015 Taylor & Francis Group, London, ISBN 978-1-138-02752-7*

# Grain refinement of Ti-6Al-4V alloy by cyclic hydrogenation and subsequent solution treatment

C.L. Huang, T.L. Chuang & Y.J. Gao
*Metal Industries Research and Development Centre, Kaohsiung, Taiwan*

T.I. Wu
*Department of Materials Engineering, Tatung University, Taipei, Taiwan*

ABSTRACT: Ti-6Al-4V alloy (Ti-64) sheet specimens were subjected to a cyclic cathodic hydrogenation and subsequent solution treatment to see the effects of processing parameters on the grain refinement of Ti-64. Electrolytic hydrogenation was performed at 50 mA/cm$^2$ for 3 hrs in 1N $H_2SO_{4(aq)}$ by adding 0.1 g/L $As_2O_3$. The subsequent solution treatment was operated in a muffle furnace at 450°C for 80 min. The microstructural changes and the hydrogen uptake were evaluated qualitatively and quantitatively by utilizing X-Ray Diffractometry (XRD: X'Pert PRO MPD, PANalytical, Netherlands, 45 kV–40 mA), Metallographic Optical Microscopy (MOM) and elemental analysis (EA: elementar Vario EL-III), respectively. The hardness variation at the near surface was measured by utilizing Akashi Vickers hardness tester (MVK-H100). The original grain size (more than 2 mm) was refined to less than 20 μm and the surface hardness was elevated from 301.6 to 478.2 VHN.

*Innovation in Design, Communication and Engineering – Meen, Prior & Lam (Eds)*
© 2015 Taylor & Francis Group, London, ISBN 978-1-138-02752-7

# Application of nano-porous-Si and trenched electrode contacts for improving the performance of poly-Si solar cells

Kuen-Hsien Wu
*Department of Electro-Optical Engineering, Southern Taiwan University of Science and Technology, Yungkang District, Tainan, Taiwan*

ABSTRACT: In this paper, trenched electrodes were proposed to enhance the short-circuit current and conversion efficiency of devices for the fabrication of NPS/poly-Si solar cells. NPS films with uniformly distributed Si nano-crystallites were first prepared on heavily doped $p^+$-type (100) poly-Si wafers by anodic etching process. Interdigitated trenches were formed in the NPS layers by Reactive-Ion-Etch (RIE) and Cr/Al metal was then deposited to fill the trenches and complete the device structures of NPS/poly-Si solar cells with Trenched Electrode Contacts (TEC's). Devices with TEC (called "TEC cells") obtained 5.5-time higher short-circuit current than that of devices with planar electrode contacts (called "non-TEC cells"). Most importantly, a TEC cell achieved 8-time higher conversion efficiency than that of a non-TEC cell. Therefore, the application of NPS surface layers with trenched electrodes is a novel approach to the development of highly efficient poly-Si solar cells.

*Applied mathematics*

*Innovation in Design, Communication and Engineering – Meen, Prior & Lam (Eds)*
© *2015 Taylor & Francis Group, London, ISBN 978-1-138-02752-7*

# Using continuous restricted Boltzmann machine to estimate the depth of sedation

Yeou-Jiunn Chen, Chai-Hong Yeng, Shih-Chung Chen, Ming-Wen Chang & Pei-Jarn Chen
*Department of Electrical Engineering, Southern Taiwan University of Science and Technology, Tainan, Taiwan*

ABSTRACT:   A proper anesthesia helps patients get through surgery without pain and avoids some other problems. By monitoring the depth of sedation, clinicians could provide a suitable amount of anesthetic administered to the patient. Then, the time needed for the recovery of consciousness and the likelihood of the patient arriving to the post-anesthesia care unit disoriented from the anesthesia would be greatly reduced. In hospitals, the bispectral index, derived substantially from bispectral electroencephalogram, is applied to monitor the depth of sedation. Generally, a patient is monitored by different types of biological systems. Therefore, to estimate the depth of sedation from vital signs can ease patient monitoring services. In this study, continuous restricted Boltzmann machines based neural network is proposed to estimate the depth of sedation. The vital signs including heart rate, blood pressure, peripheral capillary oxygen saturation, fractional anesthetic concentration, end-tidal carbon dioxide, fraction inspiration carbon dioxide, minimum alveolar concentration, and body weight are selected as the analytic features. A continuous restricted Boltzmann machine trained using the minimizing contrastive divergence is applied to extract features. Finally, a feed-forward neural network is adopted to estimate the depth of sedation. To evaluate the proposed approach, 27 subjects were asked to collect the training vital signs and leave-one-out cross-validation was used in this study. Besides, an feed-forward neural network and a modular neural network were implemented to compare the effect of features generated by using continuous restricted Boltzmann machines. The experimental results demonstrated that our approach outperformed other approaches.

*Innovation in Design, Communication and Engineering – Meen, Prior & Lam (Eds)*
*© 2015 Taylor & Francis Group, London, ISBN 978-1-138-02752-7*

# Using entropy encoding algorithm to a multilevel selection interface for BCI based spelling system

Shih-Chung Chen, Ilham A.E. Zaeni, Yeou-Jiunn Chen, Pei-Jarn Chen & Ming-Wen Chang
*Department of Electrical Engineering, Southern Taiwan University of Science and Technology, Tainan, Taiwan*

ABSTRACT:    Motor neuron disease generally weakens the muscles and results in the loss of voice and voluntary controls of the patient's limbs, hence detaching them from the outside world. Thus, the Steady State Visually Evoked Potential (SSVEP) based Brain Computer Interfaces (BCIs), which apply visual stimulus, are very suitable to play the role of communication interface for patients with neuromuscular impairments. In this study, the entropy encoding algorithm is proposed to encode the letters of multilevel selection interface for BCI based spelling systems. According to the used frequency of each letter used in spelling system, the entropy encoding algorithm is proposed to construct a variable-length tree for the letter arrangement of multilevel selection interface. Then, the Gaussian mixture models are applied to recognize the states of electrical activity of the brain. According to the recognized results, the multi-level selection interface guides the subject to spell and type the words. In order to evaluate the proposed approach, each node in tree structure of multilevel selection interface contains four sub-trees. For the root node, the letters were divided into four groups, including 26 capital letters, 26 small letters, 10 numbers (from 0 to 9), and 4 special letters (comma, dot, question mark and exclamation mark). A tree structure without considering the used frequency of each letter was implemented as baseline system. Two entropy encoding algorithms without and with dummy letter were also implemented, and denoted as EEA and EEA_D. With the consideration of the used frequency of each letter, the average length of the baseline system, EEA, and EEA_D were 3.03, 2.63, and 2.42, respectively. Therefore, the information transition rate of the proposed approach would be greatly improved at the same recognition rate of SSVEP based BCI. So, the experimental results demonstrated that the proposed approach would be useful in practical applications.

*Innovation in Design, Communication and Engineering – Meen, Prior & Lam (Eds)*
*© 2015 Taylor & Francis Group, London, ISBN 978-1-138-02752-7*

# Performance analysis of an air standard Diesel cycle with Biodiesel/Diesel blends

Jiann-Chang Lin
*Department of General Education, Transworld University, Touliu City, Taiwan*

Shuhn-Shyurng Hou
*Department of Mechanical Engineering, Kun Shan University, Tainan City, Taiwan*

Jan-Ming Lin
*Department of Education, University of Taipei, Taipei City, Taiwan*

ABSTRACT: This study aims to examine the influences of heat loss characterized by a percentage of fuel's energy on the performance of biodiesel, Diesel, and their compound in an air standard Diesel cycle. The variations in power output and thermal efficiency with compression ratio, and the relations between the power output and the thermal efficiency of the cycle are presented. The results show that the power output as well as the efficiency where maximum power output occurs will decrease with the increase of intake temperature. For a fixed maximum cycle temperature, the diesel has the largest maximum power output and the maximum power output of B20 (20% biodiesel + 80% diesel) is closed to that of biodiesel. Moreover, the relation for the efficiency is: diesel > B20 > biodiesel. The relation for the compression ratio at which the maximum power output (or maximum efficiency) occurs is: biodiesel > B20 > diesel. In other words, for biodiesel fuel, the system has to operate under higher compression ratio. The results obtained in the present study are of importance to provide a good guidance for the performance evaluation and improvement of practical Diesel engines.

*Innovation in Design, Communication and Engineering – Meen, Prior & Lam (Eds)*
*© 2015 Taylor & Francis Group, London, ISBN 978-1-138-02752-7*

# A special IR switch and Code-Maker Translator Input Device for individuals with severe disabilities

Chung-Min Wu & Jyun-Slan Liou

*Department of Computer and Communication, Kun Shan University, Tainan, Taiwan*
*Department of Electronic Engineering, Kun Shan University, Tainan, Taiwan*

ABSTRACT:   This study presents a novel Code-Maker Translator Input Device (CMTID), which enables users with severe disabilities to access the keyboard and mouse facilities of a standard personal computer, smartphone and Tablet PC via the input of suitable Morse codes processed by a Contest Fuzzy Recognition Algorithm (CFRA). The device offers users the choice of three different modes of operation, namely keyboard-mode, mouse-mode, and practice-mode. In the keyboard-mode, the user employs a special IR feedback switch to input Morse codes, then CMTID translates it into the corresponding keyboard character, symbol, or function. In the mouse-mode, the user is able to control the direction of the mouse movement and access the various mouse functions by inputting a maximum of four Morse code elements. Finally, the practice-mode employs a training environment within which the user may be trained to input Morse codes accurately and quickly via the single IR switch. Although this study presents the use of a special IR switch for the input of Morse codes, the form of the input device can be modified to suit the particular requirements of users with different degrees of physical disability. The proposed device has been tested successfully by three users with severe spinal cord injuries to operate many kinds of the application software such as generate text-based articles, send e-mails, draw pictures, browse the Internet etc. on the computer, smartphone and Tablet PC.

*Ceramics and glass*

# Effect of excess $Bi_2O_3$ content on the sintering and dielectric characteristics of 0.65 $(K_{0.5}Bi_{0.5}TiO_3)$-0.35 $BaTiO_3$ ceramic

Kin-Tak Lam
*Institute of Creative Industries Research, Xiamen Academy of Arts and Design, Fuzhou University, P.R.O.C.*

Yuan-Tai Hsieh
*Department of Electronic Engineering, Southern Taiwan University, Kaohsiung, Taiwan, R.O.C.*

Chien-Chen Diao
*Department of Electronic Engineering, Kao Yuan University, Tainan, Taiwan, R.O.C.*

Ying-Hsun Lin & Cheng-Fu Yang
*Department of Chemical and Materials Engineering, National University of Kaohsiung, Taiwan, R.O.C.*

Chin-Guo Kuo
*Department of Industrial Education, National Taiwan Normal University, Taipei, Taiwan, R.O.C.*

ABSTRACT: In this study, $K_2CO_3$, $Bi_2O_3$, $BaTiO_3$, and $TiO_2$ were mixed to form the compositions of 0.65 $K_{0.5}Bi_{0.5}TiO_3$-0.35 $BaTiO_3$ + x wt% $Bi_2O_3$ (abbreviated as 0.65 KBT-0.35 BT3-x, x = 0, 1, 2, and 3). Excess $Bi_2O_3$ was added into the 0.65 KBT-0.35 BT3 composition as the sintering aid to lower sintering temperature, to improve densification, and to compensate the vaporization of $Bi_2O_3$ during the sintering process. After being calcined at 850 °C for 2 h, 0.65 KBT-0.35 BT3-x ceramics were sintered at 1050 °C~1100 °C in air for 2 h. The X-ray diffraction pattern and scanning electronic microscope observation were used to analyze the variations of crystalline phase and microstructure of the 0.65 KBT-0.35 BT3-X ceramics. Temperature-dependent dielectric characteristics were measured with oscillating amplitude (50 mV) at 100 KHz by an HP4294A impedance analyzer by putting the sintered ceramics in a temperature-programmable testing chamber from 25 °C to 500 °C. We would well investigate the effects of sintering temperature and excess $Bi_2O_3$ content on the maximum dielectric constant and Curie temperature of the 0.65 KBT-0.35 BT3-x ceramics.

*Keywords*: excess $Bi_2O_3$ content; 0.65 $(K_{0.5}Bi_{0.5}TiO_3)$-0.35 $BaTiO_3$ ceramic; sintering and dielectric characteristics

*Innovation in Design, Communication and Engineering – Meen, Prior & Lam (Eds)*
© 2015 Taylor & Francis Group, London, ISBN 978-1-138-02752-7

# The photoluminescent properties of La$^{3+}$ ion-doped YInGe$_2$O$_7$ self-activated phosphor

Hao-Long Chen & Yeou-Yih Tsai
*Department of Electronic Engineering, Kao Yuan University, Lujhu, Kaohsiung, Taiwan*

Tai-Chen Kuo
*Department of Resources Engineering, National Cheng Kung University, Tainan, Taiwan*

Yee-Shin Chang
*Department of Electronic Engineering, National Formosa University, Huwei, Yunlin, Taiwan*

ABSTRACT:   A color-tunable La$^{3+}$ ion-doped YInGe$_2$O$_7$ self-activated phosphor is synthesized using a high energy vibrating milled solid-state reaction with metal oxides and calcined at 1200 °C for 10 hr, in air. The structure and the optical properties of (Y$_{1-x}$La$_x$)InGe$_2$O$_7$ phosphors are determined. The XRD results show that all of the diffraction peaks have the monoclinic YInGe$_2$O$_7$ phase, when the La$^{3+}$ ion concentration is increased to 3 mol%. The UV-Visible absorption spectra analysis indicates that the intensities of the absorption bands at 280–450 nm, which correspond to the oxygen deficient center of the GeO$_4$ anion, increase and then decrease as the La$^{3+}$ ion concentration increases. Under an excitation wavelength of 263 nm, the emission intensities at 487 and 545 nm, which correspond to a single ionized $V_O^{\bullet}$ defect emission, increase and then decrease as the La$^{3+}$ ion concentration is increased. The maximum intensity for these two emission peaks occurs when the La$^{3+}$ ion concentration is 0.3 mol%. This is due to lattice distortion, which induces an increase in the tensile strain. This, in turn, increases the amount of oxygen vacancies when the La$^{3+}$ ion is substituted by the Y$^{3+}$ ion. The CIE color coordinates also change from a blue → greenish → bluish region as the La$^{3+}$ ion concentration is increased from 0 to 3 mol%.

*Keywords*:   oxides; crystal growth; X-ray diffraction; optical properties

*Communication science & engineering*

*Innovation in Design, Communication and Engineering – Meen, Prior & Lam (Eds)*
*© 2015 Taylor & Francis Group, London, ISBN 978-1-138-02752-7*

# A novel Memetic Algorithm for the design of vector quantizers

Chien-Min Ou
*Department of Electronic Engineering, Chien Hsin University of Science and Technology, Taiwan*

Jih-Fu Tu
*Department of Electronic Engineering, St. John's University, Taiwan*

Chih-Yung Chen
*Department of Information Management, St. John's University, Taiwan*

ABSTRACT:   In this paper, we presented a novel Memetic Algorithm (MA) for the design of Vector Quantizers (VQs). The algorithm uses the steady-state Genetic Algorithm (GA) for the global search and C-Means algorithm for local improvement. As compared to the usual MA using the generational GA for global search, the proposed MA effectively reduces the computational time for VQ training. Numerical results show that the proposed algorithm has significantly lower CPU time over other MA counterparts running on the same genetic population size for VQ design.

*Innovation in Design, Communication and Engineering – Meen, Prior & Lam (Eds)*
*© 2015 Taylor & Francis Group, London, ISBN 978-1-138-02752-7*

# Resource allocation for higher data rate with carrier aggregation in 4G-LTE

Lu Guo & Junmin Zhou
*College of Big Data and Information Engineering, Guizhou University, Guiyang, China*

Fei Long
*College of Information Engineering, Guizhou Institute of Technology, Guiyang, China*

ABSTRACT:   In order to achieve up to 1 Gb/s peak data rate in future IMT-Advanced mobile systems, the carrier aggregation technique is introduced by the 3GPP to support very-high-data-rate transmissions over wide frequency bandwidths (e.g., up to 100 MHz) in LTE-Advanced standards. In this paper, we formulate a modified resource allocation optimization problem with carrier aggregation in Fourth Generation Long Term Evolution (4G-LTE). Utility proportional fairness resource allocation scheme has been proposed and illustrated in the past study. It will guarantee the Quality of Service (QoS), but it just discusses the situation that there are two carriers. As a result of the existence of a fragmented spectrum, we consider aggregating more than two carriers to offer a wider bandwidth and higher data rate. Therefore, we present a carrier aggregation rate allocation algorithm to allocate three carriers resources optimally among users. Finally, we prove an important theorem and the improvement of the transmission rate is verified by the simulation results.

*Innovation in Design, Communication and Engineering – Meen, Prior & Lam (Eds)*
© 2015 Taylor & Francis Group, London, ISBN 978-1-138-02752-7

# Digital business more emergency communication system based on the FPGA design and implementation

Xin Xian Rong, Luan Qiu Ping & Han Mei
*Shandong College of Electronic Technology, Zhangqiu City, Jinan, Shandong Province, China*

ABSTRACT:   This paper expounds the digital multi-service emergency communication system based on FPGA design scheme and the design process, the system can realize the video transmission via optical fiber (or wireless), such as telephone and Ethernet remote communication. The actual application topology is given. Design the network video module, telephone gateway module, Ethernet switching module, transmission module and power supply module circuit and signal process. Application of FPGA technology, selection of ALTERA corporation Cyclone II series chips, in Quartus II software platform to complete the MAC design of the key technology, Ethernet switching module. Mainly introduces the gigabit network exchange module realization process. The technical parameters in line with the national standard, and the reliability of the system are verified by simulation experiments. For the rescue and relief and field rescue and other emergency, provide the reliable and portable communication solutions.

*Innovation in Design, Communication and Engineering – Meen, Prior & Lam (Eds)*
© 2015 Taylor & Francis Group, London, ISBN 978-1-138-02752-7

# Study of the protection mechanisms on fiber transmission

Fu-Hung Shih, Luke K. Wang, Jiunn-Ru Lai & Wen-Ping Chen
*Department of Electronic Engineering, National Kaohsiung University of Applied Sciences, Kaohsiung, Taiwan*

ABSTRACT: With the exponentially increasing Internet accesses as well as the advancement of the Data Over Cable Service Interface Specification (DOCSIS) technique, CATV industry provides fiber optics connection to the Internet using cable modem and Super Multimedia On Demand (SMOD) services to increase the company's revenue. Essentially, CATV network configuration is mainly a branch topology. Fiber's breakage always brings disconnections of CATV and Internet access. Meanwhile, network service providers usually cannot be informed of network breakage instantly; network problems are being noticed only after the user's call for services. In general, network breakage location detection and repair are costly and time-consuming. Therefore, this study develops a real-time alarm system for fiber protection. The proposed system utilizes optical protection switching devices to activate a real-time routing protection switching owning to the broken fibers and display of the regions of fiber's breakage as well. The proposed system can help the network company to pin-point the breakage locations and be aware of fibers' deteriorations in order to sustain users' benefits. Both survivability and quality of service of the network are thereafter guaranteed.

*Keywords*: DOCSIS; fiber protection; real time alarm system; survivability

*Innovation in Design, Communication and Engineering – Meen, Prior & Lam (Eds)*
© 2015 Taylor & Francis Group, London, ISBN 978-1-138-02752-7

# The simulation of Solidly Mounted Resonator and filter for frequency tuning using Mason model

Kuo-Sheng Kao, Huan-Hsien Yeh & Da-Long Cheng
*Department of Computer and Communication, Shu-Te University, Kaohsiung, Taiwan*

Wei-Che Shih & Chih-Yu Wen
*Department of Electrical Engineering, National Sun Yat-Sen University, Kaohsiung, Taiwan*

ABSTRACT:   This study, focused on the simulation of an SMR filter, consists of Solidly Mounted Resonator (SMR) about various top electrode materials, thickness and ladder-type's stages. The SMR structure is constructed of a piezoelectric layer, electrodes, Bragg reflector and substrate. In order to improve the manufacturing processes, the Mason model and Butterworth-Van-Dyke (BVD) model are adopted as a simulation tool to help us maintain a competitive edge through efficient product design. According to the design rules of SMR, the bottom electrode layer is less important because of being the node in the quarter wavelength mode. The various top electrode materials with thicknesses were taken into calculation using the Mason model. Simulation result shows that the heavy mass material will cause the large resonant frequency shifting. Increasing the top electrode thickness will also increase the mass loading on SMR caused resonating frequency decreased. For the SMR filter, the amount of ladder-type stages is also investigated. The result exhibits ladder-type stages increase will increase filter's bandwidth. Besides, the Bragg reflector formed with stacking quarter wavelength thick $SiO_2/Mo$.

*Innovation in Design, Communication and Engineering – Meen, Prior & Lam (Eds)*
*© 2015 Taylor & Francis Group, London, ISBN 978-1-138-02752-7*

# The investigation of Si and Zr diffusion behaviors during the reactive diffusion—a molecular dynamics study

Jian-Ming Lu
*National Center for High-Performance Computing, Tainan, Taiwan*
*Department of Mechanical Engineering, Southern Taiwan University of Science and Technology, Tainan, Taiwan*

Tsang-Yu Wu
*Department of Mechanical and Electro-Mechanical Engineering, National Sun Yat-Sen University, Kaohsiung, Taiwan*

Jin-Yuan Hsieh
*Department of Mechanical Engineering, Minghsin University of Science and Technology, Hsinchu County, Taiwan*

Shin-Pon Ju
*Department of Mechanical and Electro-Mechanical Engineering, National Sun Yat-Sen University, Kaohsiung, Taiwan*
*Department of Medicinal and Applied Chemistry, Kaohsiung Medical University, Kaohsiung, Taiwan*

ABSTRACT: The molecular dynamics simulation was used to investigate the diffusion behaviors of Zr and Si atoms during a reactive diffusion for producing Zr silicide. The simulation results were compared with those in Roy's experimental results (Materials Chemistry and Physics 143 (2014) 1309–1314). The profiles of Mean Square Displacements (MSDs) of Zr and Si atoms at different temperatures were first used to evaluate the melting point, above which the significant inter-diffusions of Zr and Si atom occur. The diffusion coefficients near the melting point were derived by the Einstein equation from MSD profiles. On the basis of diffusion coefficients at different temperatures, the diffusion barriers of Zr and Si atoms can be calculated by the Arrhenius equation. Compared with the corresponding experimental values, the predicted diffusion barriers at the Zr/Si interface are 23 times lower than the measured values in Roy's study. The main reason is that the Zr and Si atoms within the inter-diffusion region form different local ZrSi crystal alloys in the experiment, resulting in the lower diffusion coefficients and higher diffusion barriers in the experimental observation.

*Innovation in Design, Communication and Engineering – Meen, Prior & Lam (Eds)*
© *2015 Taylor & Francis Group, London, ISBN 978-1-138-02752-7*

# High quality nitride-based material grow on silicon substrate by metal organic vapor phase epitaxy

Hon Kuan

*Department of Electro-Optical Engineering, Southern Taiwan University of Science and Technology, Yungkang District, Tainan, Taiwan*

Shoou-Jinn Chang

*Institute of Microelectronics Engineering, National Cheng Kung University, Tainan, Taiwan*

Sheng-Hsiung Chang

*Department of Electro-Optical Engineering, Southern Taiwan University of Science and Technology, Yungkang District, Tainan, Taiwan*

Chieh-Chih Huang

*Institute of Microelectronics Engineering, National Cheng Kung University, Tainan, Taiwan*

Cheng-Peng Wu, Siang-Fong Jhang & Bo-Jun Liu

*Department of Electro-Optical Engineering, Southern Taiwan University of Science and Technology, Yungkang District, Tainan, Taiwan*

ABSTRACT:  In recent years, the silicon substrate got more and more attention in the compound semiconductor field, because the silicon substrate has many advantages, such as lower cost, larger scale, and higher thermal conductivities. Therefore, we try to grow the high quality III-nitride materials on the silicon substrate in this dissertation. We use the step grading AlGaN intermediate layers to grow the high quality GaN on Si(111) substrate. The grading AlGaN can effectively release the residual stress, which was produced by the larger lattice mismatch between the GaN and Si(111) substrate. The graded AlGaN intermediate layer can also decrease the dislocation densities in the GaN epilayer.

*Innovation in Design, Communication and Engineering – Meen, Prior & Lam (Eds)*
© *2015 Taylor & Francis Group, London, ISBN 978-1-138-02752-7*

# Computational study of turbocharging an existing NA engine

Wei-Chin Chang, Yu-Hua Chien & Cho-Yu Lee
*Department of Mechanical Engineering, Southern Taiwan University of Science and Technology, Tainan, Taiwan*

ABSTRACT: In order to modify an NA engine to a turbo engine, the first step is to choose a right turbocharger. TC449 has been chosen to be the engine model. The engine model is built by using the 1D gas dynamics simulation software. Its performance is analyzed and discussed before it is modified to a boosted engine. Several parameters from the simulation results reduce the estimation and iteration process. The turbocharger is selected and the turbo engine is modeled. Its performance could be obtained from new simulation results. These results are very helpful. For example, the maximum peak cylinder pressure is relative to its mechanical strength. Thus the reliable turbo engine could be made through this computational study.

*Computer & information science*

*Innovation in Design, Communication and Engineering – Meen, Prior & Lam (Eds)*
© 2015 Taylor & Francis Group, London, ISBN 978-1-138-02752-7

# Stock selection model based on Semi-supervised Graph regularised Non-negative Matrix Factorisation, Rough Set and Grey Theory

Chuen-Jiuan Jane
*Department of Finance and Risk Management, Ling Tung University, Taiwan*

ABSTRACT:   This paper presents the consideration that the Rough Set Theory, combined with the use of Grey Prediction, Semi-supervised Graph regularised Non-negative Matrix Factorisation (SGNMF), K-means and Grey Relation, can surpass the more standard approaches that are employed in finance investment. This research focuses on the financial statement datum from the New Taiwan Economy database (TEJ) to select the optimal stock portfolio. Firstly, we collect relative financial ratio datum as the conditional attributes selection and then use GM (1,1) for predicting, SGNMF for choosing the more important conditional attributes, and Rough Set for figuring out the best portfolio. Finally, conduct fund weight distribution using the Grey relational method to reduce the investment risk. A trading system in Taiwan with good performance based on a forecasting model mentioned above is presented in this paper.

# Design and implementation of a sensor module using fiber optic power feeding for home network

Chao-Huang Wei & Chin-Chan Fang
*Department of Electrical Engineering, Southern Taiwan University of Science and Technology, Tainan, Taiwan*

ABSTRACT:    This paper describes a data transmission method and system for a sensor module via an optical fiber, wherein power and data are transmitted to the electronic system using a single optical fiber. A photovoltaic element is used to convert the light power signal into electrical energy, a portion of which can be stored in an energy storage element. The photovoltaic element can also receive or transmit a light data signal in response to a pause in the received light power signal. This technology can save additional power wiring and is easy to install sensor nodes.

*Innovation in Design, Communication and Engineering – Meen, Prior & Lam (Eds)*
© *2015 Taylor & Francis Group, London, ISBN 978-1-138-02752-7*

# Generating of visual aesthetic Fractal patterns

Guo-Fu She & Zhi-Qiang Lin
*Xiamen Academy of Arts and Design, Fuzhou University, Xiamen, Fujian, P.R. China*

ABSTRACT: This article presents the iterative function to generate the visual aesthetic fractal patterns. Based on Mandelbrot sets and the escape time algorithm on complex plane, the visual aesthetic fractal patterns are generated. The generated program development, a pictorial information system, is integrated through the application of Visual Basic programming language and development integration environment. Application of the development program, this article analyzes the shape of the fractal patterns generated by the different power orders of the Mandelbrot sets. Finally, the iterative function has been proposed as the generation tools of highly visual aesthetic fractal patterns.

*Innovation in Design, Communication and Engineering – Meen, Prior & Lam (Eds)*
© *2015 Taylor & Francis Group, London, ISBN 978-1-138-02752-7*

# Multifractal spectra of the textile pattern in quilts art design

Jun He & Artde Donald Kin-Tak Lam
*Xiamen Academy of Arts and Design, Fuzhou University, Xiamen, Fujian, P.R. China*

ABSTRACT: The design evolution of textile surface patterns is influenced by the culture backgrounds, styles, and design techniques of different times, and the pattern design of the surface is widely applied in daily life, offering unbounded pleasure to human life experience. In this work, we use the multifractal spectra to study the surface morphology of the textile pattern in quilts art design. First, a brief description of the fractal geometry is given and then, the characteristics of plane formation of the textile pattern in quilts art design are analysed. Based on box-counting method, an algorithm is derived to calculate the multifractal spectra of the plane formation of the textile pattern in quilts art design. The results of this paper include a mathematical model to describe the observation of multifractal characteristics in the plane formation of the textile pattern in quilts art design.

*Computer science & information technology*

*Innovation in Design, Communication and Engineering – Meen, Prior & Lam (Eds)*
© 2015 Taylor & Francis Group, London, ISBN 978-1-138-02752-7

# Hand Motion Identification using Independent Component Analysis of data glove and multichannel surface EMG

Pei-Jarn Chen, Ming-Wen Chang & Yi-Chun Du
*Department of Electrical Engineering, Southern Taiwan University of Science and Technology, Tainan City, Taiwan*

ABSTRACT:   This study presents an approach to identify hand motions using muscle activity separated from multichannel surface electromyogram (SEMG) and the information of data glove. There are nigh features included six features are extracted from each SEMG channel, and three features are computed from five bend sensors in the data glove. Independent Component Analysis (ICA) was used to examine its effect on independent component extraction and features reduction in this study. The results demonstrate that ICA can effectively reduce the amount of required computation data with the price of reduced identification rates. The results also indicate that the proposed method provides high accuracy (>90%) and fast processing time that is achieved to the performance of real-time system.

*Innovation in Design, Communication and Engineering – Meen, Prior & Lam (Eds)*
*© 2015 Taylor & Francis Group, London, ISBN 978-1-138-02752-7*

# An improvement to Variable-Structure Multiple-Model algorithm for maneuvering target tracking

Wenjie Zhang, Fei Long, Xia Liu & Adan Ding
*College of Big Data and Information Technology, Guizhou University, Guiyang, China*
*College of Information Engineering, Guizhou Institute of Technology, Guiyang, China*

ABSTRACT: The Variable-Structure Multiple-Model (VSMM) estimation is a popular method that is used widely in maneuvering target tracking. In this paper, based on the combination framework of Expected-Model Augmentation (EMA) algorithm and Minimal Model-Group Switching (MMGS) algorithm, two improvements will be employed for tracking strong maneuvering target. Because the model-group switching algorithm can not be activated and terminated the models without in the total model-set designed previously, model set adaptation is limited more or less. However, EMA algorithm can remedy this drawback. Thus, the combination of two algorithms will increase the estimation performance obviously. Besides, standard Kalman filter lacks adaptive ability and the transition probabilities among model set are usually invariable. So, strong tracking filter and a nonhomogeneous transition probability matrix are utilized by replacing the standard Kalman filter and transition probability matrix designed in a priori. Several simulation results used to evaluate the performance of the proposed algorithm show the effectiveness of it compared with other three algorithms.

*Innovation in Design, Communication and Engineering – Meen, Prior & Lam (Eds)*
© 2015 Taylor & Francis Group, London, ISBN 978-1-138-02752-7

# A study of genetic-based optimization of pairs-trading models

Chien-Feng Huang, Chi-Jen Hsu, Chen-An Li & Bao-Rong Chang
*Department of Computer Science and Information Engineering, National University of Kaohsiung, Taiwan, R.O.C.*

ABSTRACT: Pairs trading is an important technique that has been widely used in the real-world trading applications. Traditional approaches that solve this set of problems rely heavily on statistical methods including factor analysis and regression. In contrast to these traditional statistical approaches, recent progress in Computational Intelligence (CI) is offering alternative opportunities for solving the financial problems more effectively. In this work, we propose a CI methodology for the construction of pair-trading models using Genetic Algorithms (GA) for which reported CI-based research is sparse and lacks serious analysis. Our results showed that the proposed GA-based model is able to outperform the benchmark, and thus we expect this GA-based method to advance the current state of research in computational intelligence for financial applications and provide an alternative solution to pairs-trading for investment.

*Innovation in Design, Communication and Engineering – Meen, Prior & Lam (Eds)*
*© 2015 Taylor & Francis Group, London, ISBN 978-1-138-02752-7*

# Assessment process improvement on corroded oil pipeline by integrating web-based database system and computer aided engineering

Yu-Tuan Chou
*Department of Applied Geoinformatics, Chia Nan University of Pharmacy and Science, Tainan City, Taiwan*

Shao-Yi Hsia
*Department of Mechanical and Automation Engineering, Kao-Yuan University, Kaohsiung City, Taiwan*

Ting-Wei Chang
*Department of Applied Geoinformatics, Chia Nan University of Pharmacy and Science, Tainan City, Taiwan*

ABSTRACT:    Damage assessment of transportation components in the petrochemical industry is most important for keeping the operation and humans safe. Traditionally, the inspector derived the measured data of damaged components by non-destructive examination. Those data will be transferred to plant engineers to evaluate the lifetime using the existed specification, such as API-579. However, some flaw configuration or types are not illustrated in the assessment criteria. A computer aided analysis should be conducted to obtain more accurate stress distribution or further structural properties. Whole procedures need communication and always spend a lot of time. In this paper, an interactive website is implemented for integrating the component database and computer-aided engineering software to improve the process of corrosion assessment. The geometry and loading conditions of the component are obtained from the existing web-based database system. The corroded configuration can be derived from the collected inspection data. For simulating the stress distribution of the component structure, the commercial software ANSYS is conducted, and the simulation process is activated by APDL script language. Two case studies are examined and show the feasibility of the integration. It is promised that implementation can assist the service condition monitoring and decision system of petrochemical plant to reduce the operation cost and tracking time.

*Innovation in Design, Communication and Engineering – Meen, Prior & Lam (Eds)*
*© 2015 Taylor & Francis Group, London, ISBN 978-1-138-02752-7*

# An investigation of the relationship between performance of physical fitness and swimming ability based on the artificial neural network

Tian-Syung Lan, Pin-Chang Chen & Ying-Jia Chang
*Department of Information Management, Yu Da University of Science and Technology, Maioli County, Taiwan, R.O.C.*

Wen-Hsu Chiu
*Department of Recreational Sports Management, Yu Da University of Science and Technology, Maioli County, Taiwan, R.O.C.*

ABSTRACT:  The purpose of this study is to improve students' learning effectiveness of swimming in the physical education course, as well as to use artificial neural network to find out the connection between physical fitness and swimming ability, in order to improve the learning effectiveness of swimming teaching. This study used four indices of physical fitness, including flexibility, explosive power, muscular endurance, and cardiorespiratory endurance. The swimming test result represents the learning effectiveness of swimming teaching. This study used artificial neural network, and input the indices of physical fitness and swimming test result into Alyuda NeuroIntelligence software. After the training and prediction, the results showed that the correlation between both male and female students' cardiorespiratory and swimming ability is most significant, suggesting that cardiorespiratory endurance has a significant effect on swimming test result.

*Innovation in Design, Communication and Engineering – Meen, Prior & Lam (Eds)*
*© 2015 Taylor & Francis Group, London, ISBN 978-1-138-02752-7*

# The development of a key poses selector for use in 3D games

Feng-Ting Huang & Chien-Shun Lo
*Department of Multimedia Design, National Formosa University, Yunlin County, Taiwan*

ABSTRACT:   Characters designed to be used in a 3D game usually have several key poses, for example, standing idle, walking, running, boxing and stabbing with a sword. Heroes of a fighting game usually require fighting animation. Such animation is generally designed by animation designers with the help of motion capture equipment to record and modify the movements and the poses of which they are composed. Finally, 3D characters modeled via this process are imported into the 3D game engines for further game design. In a classical 3D game, the animation is triggered by the interface of mouse, keyboard, and joystick. However, advanced interfaces that can detect the motion of users have been developed. Interfaces such as Wii Remote and Kinect can detect the motion of users to trigger events corresponding to different animations within the game. Most games use simple motions to control the movement of the game heroes, such as hands up, waving hands etc. However, advance motion game design uses complicated motions to enable users to play the roles of the heroes. It is necessary that movements that are more complicated are recognized but the precision level of the movement must not be so precise as to make the game difficult to play. In most current games of this type, the rules of recognition are designed utilizing simple rules. However, the simplicity of the input rules results in the inability to incorporate complicated fighting motions as it is not easy to use simple rules to recognize these complex-fighting motions. Therefore, this paper proposes an automatic key motion selection and feature extraction method for developing a fighting movement database. It allows more automatic and easily generated motion recognition rules. In our designed system, Kinect is used as the interface to capture the user's motion, each movement is based around the positioning of 15 joints on the human body. A pose-based recognition system, which includes a key pose selection module and a pose matching module, is proposed. The key pose selection module finds a key pose with an Optimal Safe Differentiating Distance (OSDD). The pose matching module use the key pose and its own OSDD to identify a target animation and trigger the character's animation event in the game. This system is able to be successfully utilized within the game in real time.

*Innovation in Design, Communication and Engineering – Meen, Prior & Lam (Eds)*
© *2015 Taylor & Francis Group, London, ISBN 978-1-138-02752-7*

# User authentication of smart mobile devices and cloud storage services

Her-Tyan Yeh, Bing-Chang Chen & Ming-Chun Cheng
*Department of Information and Communication, Southern Taiwan University of Science and Technology,
Yungkang, Tainan, Taiwan*

ABSTRACT: As smart mobile devices are more common in recent years, people's lifestyles are gradually changed by the new technologies and services. Cloud storage services attract more and more attention to internet access and the trend of resource synchronization. We are able to connect to the internet via smart phones or tablet computer whenever and wherever, and fully synchronize and backup all files and document status data on different devices. However, the diverse features of smart mobile devices and the convenience of cloud storage services are double-edged swords that bring new security threats. The research simplifies the requesting process from mobile user devices to cloud storage servers, and ensures the entire service security by user identification authentication via telecommunication and the use of various encryption techniques. Mobile device users may request the cloud servers for uploading, downloading, sync and other function services. Finally, it is expected to strengthen the integrated service applications between telecommunication and cloud storage services with the research, in order to bring both sectors into the next level.

*Innovation in Design, Communication and Engineering – Meen, Prior & Lam (Eds)*
© 2015 Taylor & Francis Group, London, ISBN 978-1-138-02752-7

# Unmanned vehicle system design with multi-sensors detection capabilities

Ming-Chih Chen
*Department of Electronic Engineering, National Kaohsiung First University of Science and Technology, Kaohsiung, Taiwan*

Chia-Yen Chen
*Department of Computer Science and Information Engineering, National University of Kaohsiung, Kaohsiung, Taiwan*

Ming-Sheng Huang, Jheng-Yu Ciou & Guo-Tai Zhang
*Department of Electronic Engineering, National Kaohsiung First University of Science and Technology, Kaohsiung, Taiwan*

ABSTRACT:   This work presents a novel unmanned vehicle system, capable of detecting risks in disaster areas. The rescuer can operate an unmanned vehicle to reach the disaster area via remote control. An android-based application program is used to control the remote vehicle via wireless network. The vehicle system is capable of detecting a variety of gases by the sensors deployed on it. The gas sensors can detect methane, ethane, and carbon dioxide. After detecting the gases, the vehicle passes the collected information to the rescuer for deciding which equipments can be used in the disaster. The gas detection sensors can reduce the injury risks of rescuers before they go into the disaster area. The vehicle is also equipped with a video surveillance camera to monitor the scene and find injured persons. A robotic arm is mounted on the head of the vehicle to grip objects and gather samples. Our unmanned system can effectively reduce casualties by gathering information about the disaster beforehand.

*Innovation in Design, Communication and Engineering – Meen, Prior & Lam (Eds)*
© 2015 Taylor & Francis Group, London, ISBN 978-1-138-02752-7

# Construction of a situated-interactive learning environment for young children

Lai-Chung Lee
*Department of Interaction Design, National Taipei University of Technology, Taipei, Taiwan*

Kai-Ming Yang
*Graduate Institute of Design, National Taipei University of Technology, Taipei, Taiwan*
*Department of Information Communication, University of Kang Ning, Tainan, Taiwan*

ABSTRACT: The study aims to create a situated-interactive learning environment to benefit early stage learning for young children and tutoring for teachers so that teachers and students can play games and learning activities in an interactive environment. The innovative value of applying the interactive system is to integrate the functions in design, electromechanical and educational fields and therefore, to give young children a brand new learning experience. Conclusion is made on the basis of literature review, focus group interviews and experiments. It is listed as follows: 1. Induction of design principles on situated-interactive learning environment for young children; 2. creation of 3 scenarios focusing on situated-interactive learning games for young children; 3. establishing a situated-interactive learning area for young children. It is expected the outcome of the study can be the touchstone of integrating an interactive system with curriculums in kindergartens.

*Innovation in Design, Communication and Engineering – Meen, Prior & Lam (Eds)*
*© 2015 Taylor & Francis Group, London, ISBN 978-1-138-02752-7*

# Deployment of the Hospital Information System in medical organizations

Hou-Chaung Chen
*Orthopedics Department, Taoyuan General Hospital, Ministry of Health and Welfare, Taiwan*

E. Tungalag
*Institute of Health Care Management, Central Taiwan University of Science and Technology, Taiwan*

Ren-Hao Pan
*Department of Management Information Systems, Central Taiwan University of Science and Technology, Taiwan*

Yung-Fu Chen
*Institute of Health Care Management, Central Taiwan University of Science and Technology, Taiwan*

Kuo-An Wang
*Department of Management Information Systems, Central Taiwan University of Science and Technology, Taiwan*

ABSTRACT: It is estimated that most of the projects introducing an information an system in healthcare are not successful. Especially in developing countries the adoption rate of a health information system is quite low. Although a great number of researches and literatures have been published to address the existing problem in the area, there is a lack of focus on developing countries. This study is aimed to provide clear and concise approaches and guidance for future implementers in implementing an HIS (Hospital Information System) successfully. Divided into five distinct phases, HIS deployment methodology was developed with a list of objectives based on CSFs (Critical Success Factors) identified by reviewing the previous literatures and practical experience accumulated by the HIS implementation team at Taoyuan General Hospital (TYGH) in Taiwan. Key considerations to achieve the specific objectives in each phase were created to facilitate the readers to design and plan their own actions according to their condition and requirement. A high quality group of participants of a general hospital in Taiwan was engaged in the assessment of the designed methodology. The overall feedback we received toward the designed methodology was very positive and potential advice and recommendations by the experts were collected to make a contribution to the practice.

*Innovation in Design, Communication and Engineering – Meen, Prior & Lam (Eds)*
© 2015 Taylor & Francis Group, London, ISBN 978-1-138-02752-7

# Using Decision Tree for the analysis of National Skills Competition participants and the personality for decisions

Kung-Huang Lin

*Graduate Institute of Human Resource Management, National Changhua University of Education, Changhua, Taiwan*

Lawrence Y. Deng & Dong-Liang Lee

*Department of CSIE and IM, St. John's University, Taiwan*

You-Syun Jheng

*Department of CSIE, St. John's University, Taiwan*

Chih-Yang Chao

*Graduate Institute of Human Resource Management, National Changhua University of Education, Changhua, Taiwan*

ABSTRACT: The research utilizes t-Test and C4.5 Decision Tree to analyze commerce-related and home economics-related students. The study aims to find the pivotal personality traits which can influence the winning of a prize, by comparing the difference between the winning participants and the non-winning participants, and finally construct decision trees to achieve classification and prediction by using the Big Five personality traits to analyze the performance of senior and vocational high school students in Taiwan in the National Skills Competition. The experiment result showed that it has statistical significances on several personalities in statistical analysis between commerce-related and home economics-related students. However, because the personalities which have statistical significances are the key factor to affect whether a participant can win a prize, I go one step further to construct decision trees to predict the winner of the National Skills Competition effectively in the future.

*Keywords*: Decision Tree analysis big five personality traits; National Skills Competition; statistical analysis

*Innovation in Design, Communication and Engineering – Meen, Prior & Lam (Eds)*
© *2015 Taylor & Francis Group, London, ISBN 978-1-138-02752-7*

# Exploring users' willingness to help the online community

Li-Wen Chuang, Shu-Ping Chiu & Jun He
*Department of Digital Media Arts and Design, Fuzhou University, Xiamen, China*

Wei-Cheng Chu
*Department of Fashion Design, Shu-Te University, Kaohsiung, Taiwan*

ABSTRACT:   Social Media Communities (SMCs) are fast growing in acceptance for users, but the willingness to help online SMCs requires further exploration on what makes some online communities more successful than others. The study integrates cognitive absorption, social capital and perceived value as the antecedents of users' willingness to help the online communities; it furthermore affects the SMC continuance. The results reveal the importance of cognitive absorption and perceived value that plays a crucial role and social capital produces indirect effects in predicting the willingness to help online SMCs in the model. Based on the findings, practical implications for SNC marketing strategies and theoretical implications will be provided.

*Innovation in Design, Communication and Engineering – Meen, Prior & Lam (Eds)*
© 2015 Taylor & Francis Group, London, ISBN 978-1-138-02752-7

# Auction strategies through bidding information

Shih-Chung Chen & Chih-Hung Hsu
*Department of Electrical Engineering, Southern Taiwan University of Science and Technology, Tainan, Taiwan*

Hann-Jang Ho & Chun-Wei Ho
*Institute of Maritime Information and Technology, National Kaohsiung Marine University, Kaohsiung, Taiwan*

ABSTRACT:   Search engines are the most utilized application on the Internet. When a search engine is used for data retrieval, an inexhaustible amount of information is transmitted and spread through search results; this affects the chances of the website being viewed or clicked, thereby producing an impact on advertising effectiveness and business opportunities. Many researches study search engine keyword auctions, focusing on the impact of the keyword search auction model on Internet advertising revenue. With the second price auction as the base, this research studied a new keyword auction model in order to compare the correlation of different strategies to search engine revenue and speed of auction. This study modified AIMD, the most commonly used mechanism to control traffic on the Internet, and proposed the RAIMD auction model for individual advertisers. The pros and cons of this model were also compared to previous researches in accordance to different auction methods. Results of simulation experiments showed that RAIMD strategy is more efficient when compared to other strategies.

*Keywords*:   keyword auction; search engine; second price auction; speed of auction; auction model; RAIMD

*Innovation in Design, Communication and Engineering – Meen, Prior & Lam (Eds)*
*© 2015 Taylor & Francis Group, London, ISBN 978-1-138-02752-7*

# Design of a wireless physiological monitoring system by mesh ZigBee sensor network for hemodialysis

Ming-Jui Wu & Hsiu-Hui Lin
*Department of Internal Medicine, Kaohsiung Veterans General Hospital, Tainan Branch, Tainan, Taiwan*

Peng Yao Te
*Department of Biomedical Engineering, National Cheng Kung University, Tainan, Taiwan*

Chia-Hong Yeng & Yi-Chun Du
*Department of Electrical Engineering, Southern Taiwan University of Science and Technology, Tainan, Taiwan*

ABSTRACT: The number of patients needing dialysis increases rapidly in Taiwan. Monitoring physiological parameters during the hemodialysis period is important, but it required more and more manpower of medical care with the amount of patients. Automatic physiological monitoring for hemodialysis patients is required, but wireless monitoring is limited in broad ward and mobile beds. In this study, we developed a wireless monitoring system by mesh ZigBee sensor network to expand the coverage for long-term monitoring automatically and provide a useful user interface for nurses to manage the patient's physiological parameters. We recruited 10 hemodialysis patients who have no history of any major health problems in cardiovascularity for testing. The results demonstrate that the proposed system can effectively expand the coverage for monitoring and the accuracy of measurement has been verified by a commercial simulator (within 10%). It has high potential to be a medical system to improve the caring quality of hemodialysis patients and saving the manpower.

*Cultural & creative research*

*Innovation in Design, Communication and Engineering – Meen, Prior & Lam (Eds)*
*© 2015 Taylor & Francis Group, London, ISBN 978-1-138-02752-7*

# The need for effective cross-cultural communication in creative industries: Two case studies

Mei Zhao
*School of Journalism and Communication, Xiamen University, Xiamen, Fujian, China*

Kai Yung (Brian) Tam
*School of Cultural Industry, Xiamen University of Technology, Xiamen, Fujian, China*

ABSTRACT: As a result of globalization, people of different backgrounds constantly interact with each other. Individuals and organizations who understand the ways in which culture and communication are linked and who appreciate the range of communicative differences around the world have a greater understanding of when and why communicative misunderstandings occur and how to overcome those misunderstandings. They also recognize that interacting with people of different backgrounds brings opportunities for growth. This paper presents two cases, *Razor for the Third World* and *Reinvent the Toilet Challenge*, to demonstrate how understanding of different cultures helps industries develop new products for emerging markets and for improving the health of the world's poor.

*Innovation in Design, Communication and Engineering – Meen, Prior & Lam (Eds)*
*© 2015 Taylor & Francis Group, London, ISBN 978-1-138-02752-7*

# Social design and reflection from Hinoki Village

Yung-Chia Chiu & Shyh-Huei Hwang
*National Yunlin University of Science and Technology, Taiwan*

ABSTRACT: Cultural creative clusters get people, space, and activities together, forming a creative milieu, and leading to creative social activities and a creative economy. In Taiwan, cultural creative clusters can be divided into two types: street-type and network-type. Street-type means the cluster is formed along the street area; the network-type is formed by connecting different locations. Such clusters change social lifestyles and consumer behavior. Consumers become aware of their experiences and also like to enjoy transformations. Social design is the transformation from business creative, designing thinking, designing involvement, and design research model. Social design also combines design and sociology methods to solve social problems. Hinoki Village contains lots of unused and forgotten wooden buildings which have a century of history. To save these historical buildings, the government designs and reconstructs this village with the theme of forest cultural creativity and by promoting cultural creative tourism. This article uses semi-constructive interviews and participation-observation methods to investigate the social problems of these old buildings in Hinoki Village. We find that by support from Alisan forest railway tourism and the cultural creative industry, Hinoki Village plays a role in promoting local activity and forming a street-type cultural creative cluster. However, it looks like the financing by BOT (Build-Opera-Transfer) project results in more dominant activities in business rather than in cultural creativities. We also find a social enterprise helps to create local industry. From a social design point of view, Hinoki Village connects residents, the environment, space and the local industry, by presenting life creatively, promoting social activities, and reconstructing local environment values. Moreover, to present the concept of public, the operation of Hinoki Village intervening social development by design creativity, combining environmental education, social work and business. Thus, a new mini-community cooperation mechanism is found.

# Xiluo town community building process and analysis of development of tourism

Chia-hui Hsu & Shyh-huei Hwang
*Graduate School of Design Study, National Yunlin University of Science and Technology, Yunlin, Taiwan*

ABSTRACT: Xiluo town has rich tourist resources, and people of Xiluo make efforts to develop tourism. Xiluo Yanping Old Street Museum, founded in 2003 by Louyoung Foundation, has recorded and preserved Xiluo's culture and history for 10 years and has had a great influence on the locals. With so much effort made, tourism has not developed as well as expected. Since 2011, a severe population drain has caused the proportion of the elderly to reach 14.8% and the youth to drop from 20.7% to 16.6%[9]. Unsuccessful tourism development and lack of job opportunities are the main reason to push out the youth. This study, therefore, aims to understand the local residents' views on Xiluo development and their expectation from tourism. This study is divided into two parts. First, it analyses the views of residents on tourism developed by local organisations and residents' expectations and visions for the future development of Xiluo and tourism; second, the proposal on the future development is based on the residents' views and resources of local organisations.

The results of this study are listed as follows. When it comes to the residents' expectations for the future development, the satisfaction of daily needs and promotion of sustainability are the main concerns. Two suggestions are made on the basis of this. First, make good use of existing hardware facilities to solve the problem of childcare. Second, nurturing talents for old building repairs. As for tourism strategies, organic foods and sustainability are heavily emphasized. Organic diets, farm camps and industrial streets will be promoted. Last but not least, several tasks, based on residents' views, are suggested for the Louyoung Foundation to complete in the future. First, forge close relations among industries and promote organic foods. Second, make rural life experiences well-known to the public. Third, guide local students to highly value culture, history and old buildings. Third, form industry alliance and hold street festivals regularly as local features.

*Innovation in Design, Communication and Engineering – Meen, Prior & Lam (Eds)*
© 2015 Taylor & Francis Group, London, ISBN 978-1-138-02752-7

# The development of Chinese commercial narrative comic in the digital era

Ying-Jie Chen & Artde Donald Kin-Tak Lam
*Xiamen Academy of Arts and Design, Fuzhou University, Xiamen, Fujian, P.R. China*

ABSTRACT:   As a part of the animation industry chain, comic is the basis of animation, which is to run through the entire animation industry chain. As a result of the culture industry adapting to modern society's entertainment and commercial interests, commercial narrative comic is one of the branches of narrative comic. The development of modern digital media forms blurs boundaries between commercial narrative comic creators and readers, changes the audience's reading habits and affects the development of comics.

*Innovation in Design, Communication and Engineering – Meen, Prior & Lam (Eds)*
© 2015 Taylor & Francis Group, London, ISBN 978-1-138-02752-7

# To view interdisciplinary integration of cultural and creative industries and the experiential marketing strategy from the point of feature exhibitions—taking the exhibition experience of The Delight of Chinese Character Festival as an example

Ya-Ling Huang
*Graduate School of Visual Communication Design, Kun Shan University, Tainan, Taiwan*

Shu-Wen Yang & Hsu Fan
*Kun Shan University, Tainan, Taiwan*

ABSTRACT: This paper describes the exhibition course of participating in planning. The Delight of Chinese Character Festival, is entrusted by the Bureau of Cultural Affairs, Kaohsiung City Government. One of the major challenges is how to give consideration to the interdisciplinary integrated application of the theme-distinctive characteristic, content value and scientific media materials, as well as a plan for a marketing strategy development, based on the exhibition theme—and oriented by the marketing of the subject of cultural and creative industries. As The Delight of Chinese Character Festival is a ticketed feature exhibition, this paper, as a direct participator (curator), focuses on the development of the exhibition theme, the planning for strategic marketing, and the evaluation and appraisal of achievements after the exhibition.

*Innovation in Design, Communication and Engineering – Meen, Prior & Lam (Eds)*
*© 2015 Taylor & Francis Group, London, ISBN 978-1-138-02752-7*

# A study on the creation of animation in the first person point of view—taking the "Limitless Imagination for the Sea" of the Hongmao Harbour Cultural Park as example

Pey-Yune Hu
*Department of Motion Pictures and Video, Kun Shan University, Tainan, Taiwan*

Pei-Fang Tsai
*Department of Public Relations and Advertising, Kun Shan University, Tainan, Taiwan*

ABSTRACT:   This paper mainly studies the creation process of animation. Rather than from a traditional viewing logic, with the viewer as the third person point of view, this paper studies the creation process of "Limitless Imagination for the Sea", one of the exhibited themes of the museum in Hongmao Harbour Cultural Park of Kaohsiung City, Taiwan from the view point of the first person. The creation is based on the manifestation mode of 3D animation and themed on Hongmao Harbour, which used to be a small fishing village—and, the awe and imagination held by its residents about the sea. This paper mainly explores the development of major concepts at the early stage of creation, the relationship and imagination between the sea and nature, which is narrated by a viewer standing on a boat, with the combination of myth concepts and animation production techniques, animation manifestation in the first person and the sharing of the achievements.

*Innovation in Design, Communication and Engineering – Meen, Prior & Lam (Eds)*
*© 2015 Taylor & Francis Group, London, ISBN 978-1-138-02752-7*

# Design process analysis for lampas woven structures in textiles

Yi-Jen Guo & Wen-Dih Yeh
*Graduate Institute of Design, National Taipei University of Technology, Taipei, Taiwan*

ABSTRACT: This paper discusses the methods of lampas woven structures by analysing the process when fabrics are designed. The main feature of lampas woven structure is showing the colour weft floats to reveal the patterns on the contrasting ground, which is made by compact threads interweaving. However, it is difficult to imagine how a complex multi-layered structure can be built up by working by a 2D grid paper. The method of this study is based on the structures design logic and pattern design process to develop a novel method. The result is, this renewal design method will contribute several advantages over the woven structures in teaching and designing. By several steps, we have tried to deconstruct the original design method to simplify the processes and made the sophisticated woven design more efficient. The proposed method also enables the multi-layered structures to be applied easily in the textile design. This study shows that it is possible to inspire textile designers to be creative through logical procedures.

*Innovation in Design, Communication and Engineering – Meen, Prior & Lam (Eds)*
*© 2015 Taylor & Francis Group, London, ISBN 978-1-138-02752-7*

# Innovative design studies on modular application of traditional Kwon-Glazed Porcelain pattern

Kun Li, Ying Zeng & Can Mei
*School of Fine Art and Design, Guangzhou University, Guangzhou, China*

ABSTRACT:   Kwon-Glazed Porcelain, a traditional handicraft made in Guangzhou area, embodies the quintessence of customs and cultural inheritance in Lingnan. However, serving as ornamental porcelain and high-end household fine porcelain for a long time, it is very expensive, has a complicated manufacturing procedure and extremely specified technology. Due to that, it has become less and less popular in this rapidly developing modern society, and even the local citizens know little about it. This study tries to modularly group the patterns of Kwon-Glazed Porcelain, and summarises its traditional patterns according to compositions, styles and painting procedures. In addition, it also simplifies the lines in traditional patterns using digital technology and the procedures in composition of patterns by classification in forms of groups. As shown in this study, not only the difficulties of painting patterns are reduced by the association of digital modularisation and traditional crafts, but also a new manufacturing method is formed by the fusion of digital technology and traditional skills.

*Innovation in Design, Communication and Engineering – Meen, Prior & Lam (Eds)*
© 2015 Taylor & Francis Group, London, ISBN 978-1-138-02752-7

# The innovative design of green technology products research

Hsuan-Chu Chen
*Department of Visual Communication Design, Asia-Pacific Institute of Creativity, Taiwan*

Jui-che Tu & Shing-Sheng Guang
*Graduate School of Design Doctoral Program, National Yunlin University of Science and Technology, Yunlin, Taiwan*

Tsai-Feng Kao
*Department of Multimedia and Game Design, Overseas Chinese University, Taichung, Taiwan*

ABSTRACT: This paper presents a case study of Moka Pot through four stages. The first stage is data collecting, which extensively gathers samples of daily supplies and uses KJ method to select 20 samples. Then, natural green vocabulary and product modeling vocabulary are collected and 60 common words are obtained through a focus group. After twice vocabulary convergence, 16 Kansei words are chosen. The second stage is data analysis, which adopts factor analysis and cluster analysis of product samples and research words to generalize five representative green Kansei modeling words: "natural", "leisurely", "harmonious", "reliable" and "easy-to-use". The third stage uses morphological analysis to classify test samples into five items and four levels. A total of 1024 forms are generated and then Taguchi method is utilized to generate 16 test samples $L^{16}$ ($4^5$). The fourth stage is analysis and verification. As per S/N ratio (signal to noise ratio), this stage uses Taguchi method to find out optimal modeling scheme of each Kansei word. By virtue of modeling method of Kansei engineering, this study introduces green conceptions, adopts Taguchi quality design method, integrates innovative design approaches of green Kansei products, as well as uses Taguchi method investigate green Kansei words and main points of Kansei modeling factors. Therefore, this study can provide a reference for future research approaches.

# Creations in photography with embroidery: Taiwanese cultural values and designs

Kuo-Chun Chiu
*Department of Visual Communication Design, Kun Shan University, Tainan City, Taiwan*

ABSTRACT: The electrical properties of the poly (3,4-thylenedioxythiophene): poly (styrenesulfonate) (PEDOT: PSS) are strongly dependent on their chemical and physical structures. The mechanism of enhancement conductivity in the PEDOT: PSS films, by adding various molar concentrations of $H_2SO_4$ were further studied. The sheet resistance of the doped PEDOT: PSS film is enhanced with increasing the ratio of $H_2SO_4$, but it drops after the maximum sheet resistance. The reason for this phenomenon is that the $H_2SO_4$ reacted with sorbitol preferentially. The non-conductive anions of some $PSS^-$ were substituted by the conductive anions of $HSO_4^-$ when the residual $H_2SO_4$ reacted with PSS. After the $H_2SO_4$ doped, the sheet resistance of $H_2SO_4$-doped PEDOT: PSS film is improved by nearly 36%; the surface roughness is reduced from 1.268 nm to 0.822 nm and the transmittance is up to 91.6% in the visible wavelength range from 400 to 700 nm. However, the $H_2SO_4$-doped PEDOT: PSS films can be used as the transparent conductive electrode of optoelectronic devices in the future.

*Innovation in Design, Communication and Engineering – Meen, Prior & Lam (Eds)*
© 2015 Taylor & Francis Group, London, ISBN 978-1-138-02752-7

# Using Internet technology to raise awareness of the urban landscape skyline

Sieng-Hou Chen
*Doctoral Program, Graduate School of Design, National Yunlin University of Science and Technology, YunTech, Yunlin, Taiwan*

Li-Hsun Peng
*Department of Creative Design, National Yunlin University of Science and Technology, YunTech, Yunlin, Taiwan*

ABSTRACT: The purpose of this paper is to compare the urban landscape skyline of London, Paris and cities in Taiwan. This research suggests an efficient way to protect the landscape by using Internet technology. Many of the big cities in the world were well planned when they were founded. The urban designers considered the look of and the function of cities. However, the high speed development of society usually changes those points without much thought behind it. The coordination between modern structures and traditional architecture is very important, but the urban landscape sometimes gets dramatically transformed because of the building of skyscrapers. However despite this, the city cannot grow without constructing new structures. It is very important to preserve traditional buildings while cities continue to develop. Nevertheless, since the society changes rapidly, an integrated internet platform is thus necessary.

*Innovation in Design, Communication and Engineering – Meen, Prior & Lam (Eds)*
© 2015 Taylor & Francis Group, London, ISBN 978-1-138-02752-7

# A study on the design of inspirational products—with "Fortune Chicken" as an example

Shi-Mei Huang & Ming-Chyuan Ho
*Graduate Institute of Design, NYUST, Taiwan*

ABSTRACT: This study is composed based on the extant historical analysis, actual experiment and the probing of the importance that the modern "Fortune Chicken" merchandise has in inspiration. In this way, the crafts of the "Fortune Chicken" can be transformed into the making of a more meaningful product, and a more detailed comparison and analysis of the fabrics can be done—to investigate the application of the characteristic of inspirational goods on the "Fortune Chicken" and to zero in on the factors which will affect the design of the "Fortune Chicken" under the guidelines of creating inspirational goods. As a result, group O has the largest value among all adjectives, except relax. The result of the study provides designers a reference for creating inspirational merchandise in the future. At the same time, the "Fortune Chicken" product will be able to present its fascination, and even elevate its value and essence in the new era. Again, the "Fortune Chicken" will re-live the emotion of the bride.

*Keywords*: Fortune Chicken; product design; inspirational products

# Revitalization of China Time-honored Brand in the era of digital media

Li Shuang
*Xiamen Academy of Arts and Design, Fuzhou University, Xiamen, Fujian, P.R. China*

ABSTRACT: The analysis of the present condition of China Time-honored Brand reveals such problems of incompetence for its long history as fading impression, small audience and out-dated image. In the era of digital media today, it is the inevitable trend for China Time-honored Brand to be revitalized through digital media. Coupled with successful application of digital media to many top international brands, this essay proposes solutions to the existing problems of China Time-honored Brand through digital media such as APP, SNS and games, in the hope that it will work for the "revitalization of old firms project" launched by China Ministry of Commerce.

*Innovation in Design, Communication and Engineering – Meen, Prior & Lam (Eds)*
*© 2015 Taylor & Francis Group, London, ISBN 978-1-138-02752-7*

# Form images of local visual iconography in the genius loci

Jenn-yueh Wu
*Da-Yeh University, Changhua County, Taiwan*

Jenn-chin Lio
*Yeong-Jing Elementary School*

ABSTRACT: By bringing in the concept of visual recognition system, features of local places can be spread effectively, advantage of local places can be developed, and public recognition can be strengthened. And through the intrinsic spirit (or essence), the total images of each place are promoted. This research aims at clarifying the visual iconographies of local governments in various parts of Taiwan. Three hundred and sixty eight villages, towns, and city areas are investigated and three hundred and twenty five local visual iconographies are acquired. Besides integrating literature and data of related identity systems, four main dimensions are chosen for analysis and research. This will help further understanding the trends of local visual iconography image design. Through the values of options of colors in local iconography, material chosen in the content and dominant usage rate, the research can provide reference for related institutions in designing of local visual iconography with more flawless operating procedure and focus of theme to create truly distinctive visual iconography. Also, with integration designs of iconography, in the inner aspect, the local sense of identification is brought together and in the outer aspect, the uniqueness and competitiveness are enhanced.

*Innovation in Design, Communication and Engineering – Meen, Prior & Lam (Eds)*
© 2015 Taylor & Francis Group, London, ISBN 978-1-138-02752-7

# A study of uses and gratification of YouTube for university students

Chao-Ling Cheng & Yu-Feng Huang
*School of Journalism and Communication, Xiamen University, Xiamen, Fujian, P.R. China*

Tsung-Nan Shen
*School of Media and Design, Shanghai JiaoTong University, Shanghai, P.R. China*

Li Xue
*School of Journalism, Fudan University, Shanghai, P.R. China*

ABSTRACT: Based on the Uses and Gratification theory, through questionnaire analysis, targeting variables such as demographic, motivation and site functionality, this article aims to understand the usage of video sites such as YouTube for university students and to understand video site usages and satisfaction. The results showed that (a) the site functionality impacts user satisfaction; (b) the user motivation affects their satisfaction; and (c) demographic variables have no effect on satisfaction.

*Innovation in Design, Communication and Engineering – Meen, Prior & Lam (Eds)*
*© 2015 Taylor & Francis Group, London, ISBN 978-1-138-02752-7*

# Research on art forms of cover layouts of thread-bound books

Yan-Jun Wang & Jin-Jie Zhang
*Xiamen Academy of Arts and Design, Fuzhou University, Xiamen, Fujian, P.R. China*

ABSTRACT:    Thread-bound books originated from the Two Song Dynasty and thrived in the Ming and Qing Dynasty, they are the perfect binding form of traditional Chinese books. This paper focuses on the constitution principles and methods of cover art forms of thread-bound books, and then probes into the relation between cover components and cover beauty in form, the relation between the banding forms of thread-bound books and the cover layout forms as well as the relation between materials of thread-bound books and cover beauty. Based on the analysis on cover art forms of traditional thread-bound books, this paper puts forward the significance that modern books inherit cover designs of traditional thread-bound books, summarizes the cover design approaches of modern Chinese books under the influence of thread-bound books and points out the innovation road of book designers to push the development of modern books design and to carry forward the essence of traditional Chinese books' binding art.

*Innovation in Design, Communication and Engineering – Meen, Prior & Lam (Eds)*
© 2015 Taylor & Francis Group, London, ISBN 978-1-138-02752-7

# Research on design ideation with Taiwan cultural themes

Hsiang-Lien Lee
*Graduate Institute of Animation and Multimedia Design, National University of Tainan, Tainan, Taiwan*

Chi-Hsiung Tseng
*Graduate School of Design, National Yunlin University of Science and Technology, Yunlin, Taiwan*

Chun-Hung Liu
*Department of Visual Communication Design, Southern Taiwan University, Tainan, Taiwan*

ABSTRACT: In this study, with the perspective of design educators, this paper presents the results of determining how to incorporate Taiwanese cultural subjects into design-related knowledge, followed by forming a design strategy. A taxonomy of four synthesizing strategies was devised based on the case study: (1) a region to find the desired imagery, (2) storytelling skills, (3) stylistic skills, and (4) design practice. This strategy can be used in the early stages of the design process for novice designers. They can learn how to transmute a large amount of cultural literature into the visual design process.

# A research on the shaping of the embroidered portraits in the Qing edition of *The Romance of the Three Kingdoms*

Yu-Yu Liu
*Doctoral Program, Graduate School of Design, National Yunlin University of Science and Technology, Douliou, Yunlin, Taiwan*

Chi-Hsiung Tseng
*Graduate School of Visual Communication Design, National Yunlin University of Science and Technology, Douliou, Yunlin, Taiwan*

ABSTRACT: Taking the eight Qing illustrated editions of *The Romance of the Three Kingdoms*, 120 pieces of embroidered portraits in total, as research objects, this study uses the content analysis to discuss the shaping characteristics. (1) Features of the embroidered portraits: the feature contents include the figure proportion, facial shaping, helmet caps, costumes, accessories, boots and shoes, handheld objects and posture, and so on. It is found through arrangement that, in the aspect of the figure proportion, there are 100 six-head bodies and 94 faces with a mustache. Their costumes are of the Ming dress style, all in helmet caps to represent their identity and official positions, and mostly with waist accessories. The styles of boots and shoes include shoes and boots, and the handheld objects are usually Chinese ancient weapons. There are 114 pieces whose body postures take on the angle of 45 degrees, and 87 pieces whose head postures take on the angle of 45 degrees. The movements of their hands and feet present the basic skills and performance procedures of Peking opera. Through the above-mentioned eight items and the applications of the computer layer overlying technique, and then observing the line scribing manifestation, it is speculated that either the edition of Guanhuatang or the edition of Sanhuatang can be the block-printed one. In the edition of Baohualou, there are four embroidered portraits presenting conditions similar to right and left mirror reflection. It is inferred that it is made through slight modification after referring to several editions. (2) Features of edition: the figures in the edition of Xiaoshishanfang are of small and short proportion, while those in the edition of Shaoyeshanfang are of big and tall proportion with gorgeous shaping. The limb actions in the edition of Tongyinguan are the most abundant.

*Innovation in Design, Communication and Engineering – Meen, Prior & Lam (Eds)*
© 2015 Taylor & Francis Group, London, ISBN 978-1-138-02752-7

# A research on the manifestation relationship between the layouts of Chinese garden's artificial hills and three distances of landscape painting—taking the example of Taiwan Banqiao Lin Family Garden

Chi-Hsiung Tseng

*Graduate School of Visual Communication Design, National Yunlin University of Science and Technology, Douliou, Yunlin, Taiwan*

Pai-Chien Liu

*Doctoral Program, Graduate School of Design, National Yunlin University of Science and Technology, Douliou, Yunlin, Taiwan*

ABSTRACT:   Chinese landscape painting is the visual epitome of the life space of Chinese ancient literati as well as the specific blue print of ideal space, while gardens are like a three-dimensional landscape painting, presenting literati's insistence on ideal space. This research attempts to explore the three-remote method (loftiness, profoundness, flatness) put forward by Guo Xi, a famous pianist in North Song Dynasty, in his work The Elegance of the Bamboo and Spring—Comment on Landscape, and aims to understand the influence of Chinese landscape painting on the planning of Chinese gardens' artificial hills.

This research takes the clay artificial hills in the banyan-shade pool of Taiwan Banqiao Lin Family Garden as the research object, and considers the garden's artificial hills as a man-made space transformed twice from live-action and landscape painting. Because of the two-time transformation, the final shaping of the artificial hills is not a completely true-life presentation of the mountain scenery and landform of the hometown of Lin family in Zhangzhou. Instead, the imagination and artistic conception of the landscape architects are added into the shaping so as to satisfy people who are missing their hometown and realize personal ideal. Next, this research makes a contrast analysis between the whole layout of the artificial hills and the three-remote method, and discovers that the "loftiness, profoundness, flatness" correspond respectively to the X, Y, Z axes of the spatial three-dimensional axial directions, which is the technique paying emphasis on the spatial distance in three axial directions. On the other hand, the distance and height between the sightseeing point and the scenery in the garden is the key to presenting the three-remote method.

# Research on implementation of Hakka education by using Analytic Hierarchy Process

Tian-Syung Lan & Fu-Mei Hsu
*Department of Information Management, Yu Da University of Science and Technology, Maioli County, Taiwan, R.O.C.*

Yu-Hua Lan
*Center for General Education and Core Curriculum, Tam Kang University, New Taipei City, Taiwan, R.O.C.*

Jai-Houng Leu
*General Education Center, Yu Da University of Science and Technology, Maioli County, Taiwan, R.O.C.*

ABSTRACT:   It has been years for teaching Hakka dialect in elementary schools. There are some improvements needed to be done in schools, families and communities. It will be a great benefit for teaching Hakka dialect if some efficient solutions are provided. This study used the SWOT analysis to analyze the literatures and categorized six directions for interviewing professionals. After the interview, the results were combined as a pretest questionnaire by using Delphi method. The questionnaire was generated as a formal Analytic Hierarchy Process (AHP) questionnaire by using Likert scales to identify the factors. The participants were Hakka dialect instructors, administrative staff, government officials, and parents. This study is intended to provide objective evaluation criteria for instructors. This study can also provide proper advice for administrations, communities, and families. It can further promote Hakka dialect teaching in elementary school.

*Innovation in Design, Communication and Engineering – Meen, Prior & Lam (Eds)*
© *2015 Taylor & Francis Group, London, ISBN 978-1-138-02752-7*

# Study on the visual image of the covers of Kaohsiung Pictorial

Yi-Chen Lai
*Graduate School of Design, National Yunlin University of Science and Technology, Yunlin, Taiwan*
*Multimedia and Animation Department, TUT. Yongkang, Taiwan*

Li-Hsun Peng
*Department of Creative Design, National Yunlin University of Science and Technology, Yunlin, Taiwan*

ABSTRACT: Kaohsiung Pictorial is one of the regular propaganda pieces published by Kaohsiung City Government. With pictures and text, the construction and historical customs of the city are introduced. Consequently, the goal of this study is to analyze the images of its covers, the meaning behind their design, the relations between society and culture, and the trends of the time, by using the content analysis method. The research period was from 1980 to 2010; a total of 274 versions were checked. By virtue of the research process, the industrial and economic transformations, the political development, geological environment, customs and cultural characteristics of the city were further understood. The results showed that: (1) a cover image is an important tool for projecting impression. (2) "Place" as a subject is the main category. (3) The reports were written from the perspective of local residents and they revealed the rise of democracy. (4) Significant changeover was caused due to the parading of governing party officials. (5) The change in the words used and the styles present the transformation in social trends. (6) Labor images show the unique industrial activities and economic circumstances in Kaohsiung. (7) Various explanations of festivals and interesting happenings were made according to the historical background and the attributes of the governing party. (8) The concept of city development focused on recreation as well as the preservation and reconstruction of asset. (9) Kaohsiung Pictorial is one of the media for the Kaohsiung government to proclaim its political philosophy and governing roles.

*Keywords*: Kaohsiung Pictorial; cover design; visual image; content analysis; longitudinal research

*Innovation in Design, Communication and Engineering – Meen, Prior & Lam (Eds)*
© *2015 Taylor & Francis Group, London, ISBN 978-1-138-02752-7*

# The innovation of Chinese tea culture: From "refreshing the mind" to "tea and Chan are of one flavor"

Wei-Hsuan Fan & Huann-Ming Chou
*Department of Mechanical Engineering, Kun Shan University, Tainan, Taiwan*

Ya-Ling Huang
*Department of Visual Communication Design, Kun Shan University, Tainan, Taiwan*

ABSTRACT:   The history of tea goes back a long way in time. Originating in China, tea was first used as detoxification medicine and over time it gradually became a thirst quenching beverage. Today, tea is a spiritual drink. The moment it connected with the human mind, tea ceased to be an ordinary physical phenomenon. It has penetrated the human psyche interweaving a vibrant and unique culture. At the same time, it chronicles the course of spiritual development and refinement from "sipping tea to refresh the mind" (pin ming qing xin)" to "tea and Chan are of one flavor (cha chan yi wei)." The introduction sets forth the reasons, background, goals and procedures of this study. It gives an overview of the relationship between tea and meditative concentration and discusses how the tea-centered Chan koans demonstrate the doctrinal essence in the expression "tea and Chan are of one flavor." Literature review was used to explore the origin of tea and Chinese tea culture, the origin of Chan and the Chinese Chan School, as well as how tea culture and Chan came to integrate and culminate into the aphorism "tea and Chan are of one flavor." The conclusion: In the course of the development of tea culture, the expression "tea and Chan are of one flavor" embodies the splendor of prajna wisdom sparked by the fusion of tea culture and the Buddhist Chan School. Laymen look at this spectacular phenomenon and marvel how the merging of tea and Chan enriches the tea culture with the enigmatic flavor of Chan, whereas adepts discover the transcendental essence revealed by the aphorism. Each person sees what they are capable of seeing.

*Keywords*:   tea; Chinese tea culture; Chan; country Chan school; tea and Chan are of one flavor

*Innovation in Design, Communication and Engineering – Meen, Prior & Lam (Eds)*
© 2015 Taylor & Francis Group, London, ISBN 978-1-138-02752-7

# A research on the synergy between Buddhist Vijnana-Only Doctrine and hypnosis counseling

Ling-Tze Chao & Huann-Ming Chou
*Department of Mechanical Engineering, Kun Shan University, Tainan, Taiwan*

Ya-Ling Huang
*Department of Visual Communication Design, Kun Shan University, Tainan, Taiwan*

ABSTRACT:   The eighth vijnana, also known as the Alayavijnana, is the crux of the Buddhist Vijnana-Only Doctrine, whose central tenet "All three realms are manifestation of citta and all dharmas are vijnana-only" is founded on the existence and functions of the eighth vijnana. It exists primordially; it is immutable, and is the everlasting entity that transcends the three realms. The permanent nature of the eighth vijnana and its ability to store karmic seeds enable the process of reincarnation and the maturation of karmic retribution (vipaka) in terms of cause and condition (hetu-pratyaya). In recent decades, hypnotherapy and hypnosis counseling have become widely used as psychotherapeutic treatment and their clinical effectiveness have been recognized by therapists, psychologists, and psychiatrists. In view that the medical field of cognitive science has not seen any breakthrough for years in its study of mental functions, this paper explores the subject of mental functions from the perspective of the Buddhist Vijnana-Only Doctrine, which proposes the existence of the eighth vijnana that can store karmic seeds of past deeds. It also suggests possible solutions to overcome the research bottleneck, which could possibly diversify psychotherapeutic treatment options and hopefully transform cognitive science into an interdisciplinary study.

*Keywords*:   hypnosis; unconscious; Vijnana-Only Doctrine; Alayavijnana

*Innovation in Design, Communication and Engineering – Meen, Prior & Lam (Eds)*
*© 2015 Taylor & Francis Group, London, ISBN 978-1-138-02752-7*

# Green skylight: An overview of the aesthetics and eco-friendliness of a courtyard

Pei-Ying Ou
*Department of Mechanical Engineering, Kun Shan University, Tainan, Taiwan*

Ya-Ling Huang
*Department of Visual Communication Design, Kun Shan University, Tainan, Taiwan*

Huann-Ming Chou
*Department of Mechanical Engineering, Kun Shan University, Tainan, Taiwan*

ABSTRACT:    The courtyard is an architectural form commonly seen in traditional Chinese architecture. This paper explores its applications of as well as the various ideas related to it. Specifically, this paper studies the structural forms and functions of courtyards in traditional architecture, its cultural significance, and also the green concepts embedded in its design. By increasing the understanding of and appreciating the aesthetics and eco-friendliness of a courtyard, it is hoped that this unique structure can be widely used in modern architecture to inspire innovative thinking and applications in modern ecological construction and green living.

*Keywords*:    courtyard; green concept; ecological construction

*Innovation in Design, Communication and Engineering – Meen, Prior & Lam (Eds)*
© 2015 Taylor & Francis Group, London, ISBN 978-1-138-02752-7

# The bonding of popular topics with grassroots economy in Taiwan: A case study using the Rubber Duck sensation

Lan-Kuo Sung
*Department of Mechanical Engineering, Kun Shan University, Tainan, Taiwan*

Ya-Ling Huang
*Department of Visual Communication Design, Kun Shan University, Tainan, Taiwan*

Huann-Ming Chou
*Department of Mechanical Engineering, Kun Shan University, Tainan, Taiwan*

ABSTRACT: New discussion topics crop up almost every day in our lives. But among them, which ones will draw the attention of the populace and turn into an ongoing hot topic and which ones are able to sustain themselves by successfully stimulating economic activities? This study uses the "Rubber Duck," a pop culture sensation, by content analysis to explore the bonding of popular topics with grassroots economy in Taiwan.

*Keywords*: popular topic; grassroots economy; mass media

*Innovation in Design, Communication and Engineering – Meen, Prior & Lam (Eds)*
*© 2015 Taylor & Francis Group, London, ISBN 978-1-138-02752-7*

# The assessment of the impact of character modeling on design fixation in movement expression of character animation

Huang-Yao Lin
*Department of Digital Media Design and Management, Far East University, Tainan, Taiwan*
*Doctoral Program, Graduate School of Design, National Yunlin University of Science and Technology,*
*YunTech, Yunlin, Taiwan*

Li-Hsun Peng
*Department of Creative Design, National Yunlin University of Science and Technology, YunTech, Yunlin, Taiwan*

ABSTRACT: Creative movement expression is relatively important for character animation. However, in defining the external imagery of animation characters, design fixation possibly happens, so that the movement and physical performance of character animation may affect dynamic animation operation. In this study, 60 students learning dynamic animation performance are the subjects, among whom 40 have operation experience of character animation for more than one year. All of them are asked to create a 48-sec./frame basic character walking in a separate operating environment and divided into two groups in the same situation: one is given a basic character modeling with constraints of stature, gender, and age; the other is given the character modeling of geometric prototypes. Modeling skeletons, bind skins, and controllers are the same components in the two groups. After the performing data are collected, the principle of animation is adopted as the basis for this assessment of movement. The result shows that the movement production implied with character modeling has more movement performance of actual characters so as to exhibit specific characteristics; however, due to the limitation of character modeling, the movement performance is stereotyped. By the result of this study, the implied message controlling is assessed in the animation production team during the dynamic creation process; and further, how animators induce the proper and actual movement of characters in producing the animation.

*Innovation in Design, Communication and Engineering – Meen, Prior & Lam (Eds)*
© 2015 Taylor & Francis Group, London, ISBN 978-1-138-02752-7

# Value-added application for e-learning into Chanel's on-line exhibition through internet technology

Li-Hsun Peng
*Department of Creative Design, National Yunlin University of Science and Technology, Taiwan*

Chia-Hsin Hsueh
*Doctoral Program, Graduate School of Design, National Yunlin University of Science and Technology, Taiwan*

ABSTRACT:    With the transition of culture, widely adapted Capitalism, and the critical influence of the mass media, fashion design gradually has become a trend in the postmodernera that can be understood by a global audience. As we know, Coco Chanel is one of the icons of fashion design, especially known for her famous "little black jacket". This study primarily analyzes the factors influencing Chanel's brand image towards the potential contribution of value-added applications for e-learning in "*Chanel the Little Black Jacket online Exhibition*". This exhibition shows the astounding versatility of Karl Lagerfeld and Carine Roitfeld's reinterpretation of Chanel's iconic little black jacket. The study will be using phenomenography and case study as the main methodologies. It analyzes the relationships between the internet technology and value-added applications for e-learning from design, culture, and consumption perspectives through related documents. Besides, this research realizes the in-depth implication and brand recognition of digital archive through visual literacy. Finally, we not only hope the explanations will provide a method design approach for the e-learning development of value-added applications with fashion aspects for future research and design references, but also ensures Chanel's future as a timeless classic.

*Innovation in Design, Communication and Engineering – Meen, Prior & Lam (Eds)*
*© 2015 Taylor & Francis Group, London, ISBN 978-1-138-02752-7*

# The heritage and innovation of the graceful cheongsam culture: A case study of Taiwanese designer labels

Hsi-Hwei Lo
*Department of Mechanical Engineering, Kun Shan University, Tainan, Taiwan*

Mei-Tuan Tsai
*Center for General Education, Kun Shan University, Tainan, Taiwan*

Huann-Ming Chou
*Department of Mechanical Engineering, Kun Shan University, Tainan, Taiwan*

ABSTRACT: Though not many Taiwanese fashion brands excel in traditional Chinese design, their unique and distinctive style nonetheless enjoys enduring popularity. The apparel stores in the business area near Taipei's Yung Kang Street, for instance, showcase designer labels that integrate traditional design elements with cultural and artistic sensibility, as well as the traditional Chinese cheongsams (qipao) that flaunt the sumptuous style of old Shanghai. Xu Rongyi and Xu Yikao are dressmakers who specialize in hand-sewn Shanghai-style cheongsam. They have witnessed the bloom of this style in Taiwan during the 1940's and 1960's and its decline after the 1970s when local designers sprung up, one after another, to establish their own labels. Some of them, such as Tsai Mon-Sha's LONDEE (1976), Lee Chun-Chih's Greenleaf (1976), SHIATZY CHEN (1982), as well as Nadia Lin (1988) and Chang Chen's (2000) eponymous brands, are now celebrated designer labels. Combining elegance and modern chic and infused with the quintessence of Chinese culture, the modern cheongsam's classy style is perfect for important occasions. Insisting on the exquisite quality of handmade craftsmanship, today's cheongsam has been marketed as a product of cultural innovation using sales platforms that employ visual media. Innovations and insistence on quality adapt modern cheongsam to the latest fashion taste and psychology and at the same time satisfy the demand of formal occasion dress. In the competitive designer apparel market, cheongsam continues to attract loyal customers and sustain the legacy of our graceful "national dress."

*Keywords*: cheongsam (qipao); fashion designer; Chinese element (oriental element); market; brand

# Innovation and environmental protection in Taiwan's "vegetarian diet culture"—survey targeting religious groups

Chin-Yuan Lin & Huann-Ming Chou
*Department of Mechanical Engineering, Kun Shan University, Tainan, Taiwan*

ABSTRACT:  Due to the environment protection trend in recent years, global warming and other related issues have become significant topics. In response to that, energy–saving and carbon-dioxide-reduction were undertaken on a worldwide scale, and actions such as more-veg-less-meat, while the UN also recommend vegetarian to prevent global warming. Stop eating meat, cycle more and consume less will help to delay the global warming. Even though the vegetarian population is growing notably, among them, mostly are elderly or have a religious purpose. Yet it is still not commonly accepted because conventional meat dieters are rejecting vegetarianism somehow for its flavor and taste. If we can stress and specify the advantages of vegetarianism, this will encourage more people to join. As a matter of fact, the key point to invite more people to adopt vegetarianism is the appetite, through innovation, to optimize the food delicacy, to improve the taste and presentation will definitely provide an irresistible attraction to those who were against it. By exploring the environmentally friendly vegetarian culture, we look forward to this positive discussion that will lead to more understanding of the benefits from vegetarianism, which will furthermore turn the stereotype against it, and eventually have it become a mainstream diet. To change from carnivore to vegetarian will not only improve mankind's health, but also comply with human being's effort and attention toward the ecology and environment, to avoid any harm to the motherland, to stay ahead of food shortage, to allow Earth to breathe freely and to let all beings live as naturally as they should forever. It is a substantial subject to promote Taiwan's vegetarian culture; it shall be done by means of innovation and creation. The article hereby will point out further research of Taiwan's vegetarian culture related with religious groups, and meanwhile, it will analyze and put together all these efforts of vegetarian culture promotion and innovation made by Taiwan's religious groups.

*Keywords*:  vegetarian diet; cooking culture; environmental protection; religion

*Innovation in Design, Communication and Engineering – Meen, Prior & Lam (Eds)*
*© 2015 Taylor & Francis Group, London, ISBN 978-1-138-02752-7*

# A study of the combination of traditional craft and modern design in bamboo furniture

Shih-Hsing Wu
*Doctoral School of Design, National Yunlin University of Science and Technology, Douliu, Yunlin, Taiwan*

Ming-Chyuan Ho
*Department of Industrial Design, National Yunlin University of Science and Technology, Douliu, Yunlin, Taiwan*

ABSTRACT:   Bamboo furniture used to be a specialty of Taiwan, where the art of bamboo craft featured daily lifestyle goods. During the period of Taiwan's soaring economy in 1970 and 1980, there were about 2000 bamboo factories in Chushan, but the bamboo industry has declined over the years. Since the 21st century, people have begun to care more and more about environmental protection and emphasized the importance of using green materials. Therefore, bamboo has become a popular material, and the value of bamboo furniture has been elevated. However, there is little research on the traditional craft of bamboo furniture. Hence, this study aims to explore: 1. The style, production process, and application of traditional craft in bamboo furniture; 2. The style, production process, and application of modern design in bamboo furniture; 3. The practice of combining traditional craft and modern design in bamboo furniture. Based on grounded theory, this study explores the above mentioned issues using action research and case study methods. Three famous bamboo furniture companies were interviewed in order to understand the interaction between traditional bamboo furniture and modern design. The result shows: In order to survive in the future, bamboo furniture will need to combine traditional craft culture and techniques with the concepts of modern design in practice. In this way, bamboo furniture can regain its popularity and value in the eyes of the consumer.

*Innovation in Design, Communication and Engineering – Meen, Prior & Lam (Eds)*
© *2015 Taylor & Francis Group, London, ISBN 978-1-138-02752-7*

# Using traditional materials to revitalizing East Timor's cultural industries

Chi-Yu Pan
*Doctoral Program, Graduate School of Design, National Yunlin University of Science and Technology, YunTech, Yunlin, Taiwan*

Li-Hsun Peng
*Department of Creative Design, National Yunlin University of Science and Technology, YunTech, Yunlin, Taiwan*

ABSTRACT: In the past ten years, Cultural Industry and Creative Economy were getting more and more important with the development in many countries. Through this research of local materials, cultural knowledge and traditional intelligence to produce attractive products and services, not only could increase the economic growth but also will maintain our traditional culture for the next generation. Nowadays, creative design also becomes a key strategy and major issue for indigenous groups. Since the independence in 2002, East Timor has come through a hard way to form a country. Having undergone Portuguese colonialism and then the domination of Indonesia, East Timor is trying to find its own way to rebuild its culture through creative design and through using the traditional materials. This paper aims to explore cultural creativity and revitalization to help the local industries facing the tough challenges. How the people take creative design as a basis to help themselves in the economical, environmental, social, and cultural domains. In the case study, a researcher visited East Timor to understand the development of the local industry. This research, targeting at local indigenous groups and their projects, we expect to sum up their experience from the transformation of local industries. Through the discussion of creative design and materials sciences theories in the research, we hope that people could preserve their traditional cultural elements to develop their own culture. This will not only create opportunities, but also promote the upgrading of the local industry. We expect that the research will help to set up models and the development of traditional materials of creative design in East Timor.

*Innovation in Design, Communication and Engineering – Meen, Prior & Lam (Eds)*
*© 2015 Taylor & Francis Group, London, ISBN 978-1-138-02752-7*

# Golden puppet shows elegance reproduction in the digital era

Shih-Mo Tseng
*Full Graduate School of Design Doctoral Program, National Yunlin University of Science and Technology, Yunlin, Taiwan*

Chao-Ming Wang
*Department of Digital Media Design, National Yunlin University of Science and Technology, Yunlin, Taiwan*

ABSTRACT: Puppet shows in Taiwan have been circulating for over 200 years. Golden puppet shows led to a peak in the 1960s. Subsequently, although the TV puppet shows have combined puppet shows and wireless media successfully and created another peak in the 1970s, its performance types are no longer in contact and interaction with the audience directly. In this study, we have produced digitized sound and light effects combined with the LCD screen, high-fidelity speakers and projection technology. Better than ever to make visual and hearing effects, applied to public performances puppet shows scene. It applies to public puppet shows performance opportunities. Oral and operating puppets skills combine perfectly to attract audience attention and thoughts. We over-rely on the performance of electronic device auxiliaries. Prolonged operation of the device must be overcome easily and generate system errors and problems of the program itself must not be faulty. It is done through the opportunities of large-scale performance to find and to improve problems, and achieve the goal of perfect performance gradually.

*Keywords*: puppet shows; interactive device; digital technology; interactive performances

*Innovation in Design, Communication and Engineering – Meen, Prior & Lam (Eds)*
© 2015 Taylor & Francis Group, London, ISBN 978-1-138-02752-7

# The era of miniature cultural and creative industry: Integrating style shaping with business pattern into brand construction

Tsen-Yao Chang

*Department of Creative Design, National Yunlin University of Science and Technology, Yunlin, Taiwan*

Wen-Shin Chang

*Department of Industrial and Commercial Design, National Taiwan University of Science and Technology, Taipei, Taiwan*

ABSTRACT: The industrial pattern of Taiwan has considerably expanded since its shift from the Original Equipment Manufacturer (OEM) phase. Given the rise of the cultural creative industry and the corresponding improvement in life aesthetics at the national level, the establishment of a miniature cultural and creative industry welcomes an open mode of management, in which brand core values are developed to the greatest degree by using the lowest amount of capital and the lowest number of employees. Despite the small scale of this new industry, its creativity, autonomy, and ideal characteristics have motivated entrepreneurs to engage in business, thereby enhancing the creativity and diversity of the whole market. In this work, we discuss the miniature cultural and creative brand by performing a two-step in-depth interview, keyword clustering, root analysis, and other stages of study to structure the style shaping and business pattern of the miniature cultural and creative industry. At the first stage, the keywords of the miniature cultural and creative industry were collected through interviews with three entrepreneurs from the north of Taiwan. The interview outline was then modified for the interview in the next stage. During the second stage, the interview object was expanded to the whole Taiwan, including seven entrepreneurs from New Taipei City, Hualien county, Tainan, and Kaohsiung, to enhance the diversity of the samples and to identify any differences in the miniature cultural and creative industry across these areas. The interview content was also processed via grounded coding. The secondary data collected in the first stage were then combined with the analyzed and concluded information in all stages to understand the business pattern of the miniature cultural and creative industry as well as to determine how the style of the brand could be shaped. Shaping the style of the brand in the miniature cultural and creative industry required four influential factors, namely, emotional motivation, cultural background, life experiences, and vision concept. As revealed by our findings, the core value of the micro brand is to deliver a lifestyle with an emotional aspect and vision concept that could motivate the founders of the brand. We also found that life experience and cultural background play important roles in creating the lifestyle that makes up the annular business pattern of the miniature cultural and creative brand focused on emotional value.

# Integrating service-oriented design for innovative course model construction

Tsen-Yao Chang
*Department of Creative Design, National Yunlin University of Science and Technology, Yunlin, Taiwan*

Miao-Chen Lin
*Department of Industrial and Commercial Design, National Taiwan University of Science and Technology, Taipei, Taiwan*

ABSTRACT: In this age when culture, innovation, and life industry are closely connected with "design", integrative and creative personnel who are industry oriented must be developed to bring together the academy and the industry, introduce new vigor into the cultural, innovative, and design industry, and satisfy the industrial requirement. This research focused mainly on how to introduce service design thinking into the teaching of integrated innovative design and attempted to apply the systematic and procedural ideology of service design into courses that can promote the ability of students to theorize, design, and develop a feasible course model. Based on actual design practice and service design thinking and tools, students explored the on-site situation at the "Sun Link Sea Forest Ecological Holiday Park" and applied integrated design thinking and methods, such as the experience service design. The students observed and worked independently, while they enhanced their empathy and integrative design ability, gained some insights into the problems, and obtained a solution to rebuild the specific situation. According to the action research theory, the operational process of the course can be divided into four parts: plan, action, observation, and reflection. By non-participant observation and analysis of coded data collected using questionnaires, this paper determines the effects of service design tools on students at different stages of the study. Results suggested two main domains, namely, cognitive and emotional domains that contained nine index abilities: fluency, creativity, flexibility, accuracy, and sensitivity for the cognitive domain and risk-taking tendency, challenge-taking tendency, curiosity, and imagination for the emotional domain. All these index abilities are necessary to perform the integrated design. Finally, this research examined whether the course arrangement can satisfy the ability training for the program and comprehensively developed a course model in introducing service design into integrated and innovative teaching, which can be a reference for future course planning among integrated design and cultivating personnel with relevant abilities in the present cultural and design industry.

*Innovation in Design, Communication and Engineering – Meen, Prior & Lam (Eds)*
*© 2015 Taylor & Francis Group, London, ISBN 978-1-138-02752-7*

# Implanting the value of service design into the development of religious cultural and creative merchandise

Tsen-Yao Chang

*Department of Creative Design, National Yunlin University of Science and Technology, Yunlin, Taiwan*

ABSTRACT:   Taiwan is a place where religious diversity is accepted. In terms of regional management, along with a change in tourism patterns, unique features such as the belief, history, culture, and art in the religious cultural field have become the advantages for in-depth travel development. In the management of religious cultural originality fields, ideology passing, and blessing seeking, a contact point needs to be created for product development, marketing, and gift-giving to efficiently spread religious characteristics. Therefore, by introducing the service design thinking and methods into the field of religious culture, this study explored the market demands and provided feasible suggestions for service design, thereby eventually reflecting ways of creating new values through service design as well as developing new thinking and models to create new products. In the study on the Kong Fan Temple in MaiLiao village, the initial part of the research design utilized a field investigation to learn about the current state of the palace, which was then followed by a non-participant observation on the arriving and departing tourists. Next, an analysis was conducted by using service design tools to illustrate the customer journey map of the target field and consequently determine the various insufficiencies of the field. During the second stage, the qualitative data codes gained from the service design procedure were combined with the documentary data to conduct a comprehensive analysis and formulate a design process for cultural products. Afterward, a positive analysis was performed to induce the developing model for the products with religious cultural originality from advanced practice. Finally, through an interview with the Kong Fan Temple committee, both designers and sellers were considered in the endeavor to modify and propose "the Introduction model of service design for products with religious cultural originality." Results can be presented to managers of religious cultural originality fields, to help the managers apply the service design philosophy, which aids in stimulating innovation and exploiting opportunities for new product developments. Through the application of the service design thinking, a lasting regional advance would hopefully be enhanced by listing religious cultural fields as the foundation of regional tourism.

*Innovation in Design, Communication and Engineering – Meen, Prior & Lam (Eds)*
© 2015 Taylor & Francis Group, London, ISBN 978-1-138-02752-7

# Add brand value to traditional enterprises through design using the package design of pickled products at Miaoli area as an example

Chien-Cheng Chang, Cheng-Yuan Lin & Tzu-Heng Ku
*Department of Industrial Design, National United University, Miaoli, Taiwan*

ABSTRACT: Under the competitive marketing commercial environment, branding and design are keys to the enterprise competition. How to add value to a brand by design is a goal and challenge to every enterprise. In this study, expert interviews and the focus group method are used to explore the current status of pickled product industry at Taiwan Miaoli area and the customer's needs and viewpoints of the package design of pickled products. Then, proper brand strategy and design specifications are built upon brand positioning result so as to design and develop a new series of package design. In this study, the Pickle Museum is used as an example for brand strategy and package design. Managers of major local brands of pickled products are interviewed. Furthermore, a total of 100 focus group subjects are invited for a marketing survey and semantic differential test. From the study results, specific styles for the Pickle Museum are identified. Moreover, the market study helps set up a design strategy for the new product design and development. The theoretical design and development model can serve as a reference for traditional enterprises to add value to their brands. Through expert diagnosis and market positioning as well as cultural creative concepts, the market needs and brand image perceptions can be transformed into product design specifications so as to reduce the risk of the new product development.

*Keywords*: brand strategy; focus group; market positioning; Hakka culture

*Innovation in Design, Communication and Engineering – Meen, Prior & Lam (Eds)*
© 2015 Taylor & Francis Group, London, ISBN 978-1-138-02752-7

# Breaking through the ontological confines of biological materialism in cognitive science

Jui-Wen Yu & Huann-Ming Chou
*Department of Mechanical Engineering, Kun Shan University, Tainan, Taiwan*

ABSTRACT: Cognitive science has successfully broken down the barriers between biology, physics, neuroscience, psychology, and anthropology through "interdisciplinary integration," a pioneering approach that adopts integrative thinking and research methods. Cognitive science studies mind-related phenomena, an area that has been long ignored by physical sciences but is always at the heart of humanities, arts and social sciences. Unfortunately, since it grounds its study of the mind in mental-material monism, cognitive science ascribes all mental functions to physiological phenomena produced by the "brain map" and is essentially jumping into the quandary of biological materialism. Without positive ontology as its theoretical basis to adequately account for the ontological entity that underlies the neuroplasticity achieved by genetic transcription, cognitive science will either box itself in the confines of biological materialism or be besieged by all sorts of irresolvable philosophical paradoxes. This article proposes the eighth vijnana of Buddhism to be the positivist ontological entity of the mind. The eighth vijnana is neither mind nor material, yet can give rise to both as well as all mental functions. In other words, it is the basis of our mental schema. Moreover, since the existence of the eighth vijnana can be realized and verified by each and every person, its nature is in conformance with the positivist spirit of science. It is hoped that the understanding of the eighth vijnana can provide vital indications for research and rethinking in cognitive science, which, given its interdisciplinary spirit, should be receptive to any suggestions conducive to its research development.

*Keywords:* cognitive science; ontology; the eighth vijnana

*Innovation in Design, Communication and Engineering – Meen, Prior & Lam (Eds)*
© *2015 Taylor & Francis Group, London, ISBN 978-1-138-02752-7*

# The possibility of cultural proximity between Taiwan, Singapore and Japan in Kawaii culture

Chu-Yu Christine Cheng
*Department of Creative Product Design, Southern Taiwan University of Science and Technology, Taiwan*

Hung-Yuan Chen
*Department of Visual Communication Design, Southern Taiwan University of Science and Technology, Taiwan*

ABSTRACT: The study of "Kawaii culture" could be considered an important research that helps clarify Japanese culture and its globalization role. Kawaii (which means, cute™ in English), is one of the most important components in Japanese popular culture. It has been spread throughout the world and became a transnational culture and aesthetic for youth around the world. This trend could be traced back to the millennium year in Taiwan and Singapore's history. A few social academics were concerned the strong globalization trend might have affected the youth's national identity or, it was just cultural proximity in the Asia region. Nevertheless, it has been almost one and a half decade, and more and more social communication networks reinforce the Globalization significance. However, I argued the particular connection of "geo-cultural regions" and nostalgia notion from colonial history; would the globalization overtake cultural proximity so easily? Nowadays, we may be able to get a chance to redefine the influence cause of Japanese culture in Taiwan and Singapore again.

*Innovation in Design, Communication and Engineering – Meen, Prior & Lam (Eds)*
© 2015 Taylor & Francis Group, London, ISBN 978-1-138-02752-7

# An exploration of documentary film recommender systems

Yu-Feng Huang
*School of Journalism and Communication, Xiamen University, Xiamen, Fujian, P.R. China*

ABSTRACT: With the combination of technology and life increasing on the basis of the mobile platforms, a growing number of accesses to information channels were provided for the general public. Networks formed by a variety of user-centric social media became more compact, this phenomenon allows the audience more choice in terms of documentary qualities. Unlike the passive recommendations in the past, this paper borrows the market operation point of view and data mining technologies of the age of the focus media and put forward a proactive recommendation system. First, analyze the users with the need for documentaries through the dialog boxes of various social networking sites. Then, through Fuzzy Clustering Theory, surmise the user demand for Chinese documentaries, and actively recommend suitable Chinese documentaries in social networks. Finally, friends of friends by the users of social media create a system that spreads like diffusion. This paper proposes an Active Recommendation System and got satisfactory results from actual verification and analysis of satisfaction.

*Innovation in Design, Communication and Engineering – Meen, Prior & Lam (Eds)*
© 2015 Taylor & Francis Group, London, ISBN 978-1-138-02752-7

# From industrial heritage regeneration to regional revitalization—a case study of Checheng Timber Industry Museum Park in Taiwan

Chuan-Jen Sun
*Graduate School of Design, National Yunlin University of Science and Technology, Douliu, Yunlin, Taiwan*
*Department of Creative Public Communication Design, TransWorld University, Douliu, Yunlin, Taiwan*

Shang-Chia Chiou
*Graduate School of Design, National Yunlin University of Science and Technology, Douliu, Yunlin, Taiwan*

ABSTRACT:    An industrial heritage site is the best witness of the growth and decline of a region's economy and the alternation of the industries. It carries not only historical value and significance of a region's development but also the collective memories shared by people of this region. Communities that were established in the early days due to industries often experienced a lagging development and even population decline following the discontinuance or abandonment of the industries they relied on. An importance of conserving an industrial heritage site is to highlight the distinctive characteristics of the community and make it an integral part of the local history and culture. In recent years, Taiwan has devoted much effort in preserving and reusing industrial heritage sites. However, in most cases, the "sites" were preserved but not the "heritage." As a result, some of these industrial heritage sites became commercialized and lost their original characteristics and cultural significance. Therefore, this study aims to examine the impact of the regeneration of Checheng Timber Industry Museum Park, an industrial heritage site in Taiwan, and how it revitalized local tourism and became the driving force behind community development and the success of creating a sense of community. Furthermore, the study will also analyze its preserving value and the possibility of constructing an industrial heritage network.

*Design theory & knowledge innovation*

*Innovation in Design, Communication and Engineering – Meen, Prior & Lam (Eds)*
© *2015 Taylor & Francis Group, London, ISBN 978-1-138-02752-7*

# A framework of teaching preparation and planning to arouse children's creativity through the colour-based story editing approach

Yu-Hsien Wu & Hsien-Jung Wu

*Department of Commercial Design, National Taichung University of Science and Technology, Taiwan*

ABSTRACT: Children usually recognize daily life objects based on colour experience such as a yellow taxi, white milk and green leaves. Colour recognition plays an important role in learning objects and the environment for children. Therefore, learning and recognizing colors from a parent's or teacher's instruction has a great influence on children's recognition ability and imagination, which is "the beginning of creation" (George Bernard Shaw). A one-way teaching method based on adults' colour experience possibly will restrain children's creativity.

To avoid the above-mentioned problem, this research proposes a systematic approach using story editing to assist teachers or parents in preparing and planning instructions when they teach children how to recognize colours and objects. This approach includes a four-module framework: information delivery module, teaching module, evaluation module and feedback module. It provides instructors with a guidebook of pre-teaching preparation, instruction implementation and post-teaching analysis. The advantage of the story-editing approach is to integrate the senses of vision and hearing, which guides children to correlate characters with respective colours and enhance their imagination.

This systematic framework of teaching preparation and planning is especially useful for teachers or parents to guide young learners who use the colour experience they learn from adults to recognize objects and the environment. Its interactive and multi-modal manner provides instructors with a hands-on sample to arouse children's creativity.

Innovation in Design, Communication and Engineering – Meen, Prior & Lam (Eds)
© 2015 Taylor & Francis Group, London, ISBN 978-1-138-02752-7

# Ontological analysis on the 3D animation techniques from a perspective of "Art Automata"—exemplified by a reference to traditional Quanzhou marionette performances

Chaomiao Chen
*School of Fine Arts and Design, Quanzhou Normal University, Quanzhou, Fujian, China*

ABSTRACT:    Based on the theories of "Art Automata" and "Patterns and Noise", this paper finds a good combination of both modern technology and traditional culture by an ontological analysis on the 3D animation techniques. It studies Quanzhou marionette performances, a typical local folk art, and analyzes its styles, especially focusing on the balance between automata and control as well as the balance between patterns and noise in the performances. Through a discussion about the experimental agreement of theory and practice, this paper better demonstrates the relations between 3D animation techniques and animation styles.

*Innovation in Design, Communication and Engineering – Meen, Prior & Lam (Eds)*
© *2015 Taylor & Francis Group, London, ISBN 978-1-138-02752-7*

# Design and development strategy of lightweight product packaging in green life cycle

Hong-Yi Chen & Jui-Che Tu
*Graduate School of Design Doctoral Program, National Yunlin University of Science and Technology, Douliou, Yunlin, Taiwan*

Chuan-Ying Hsu
*Department of Business Administration, Dayeh University, Taichung, Taiwan*

ABSTRACT: The promotion and implementation of lightweight product packaging in industrial, official and academia for years have drawn the public's attention on environmental issues. According to a policy survey in May 2013, people could accept and want to purchase goods with environmental packaging. Although the effectiveness of the policy was significant, different enterprises have different interpretations and understanding of the concept of lightweight product packaging. Hence, actual lightweight product packaging designed by different enterprises varies widely. Consumers tend to make the wrong selection while doing purchases. This study aims to conduct a review of literature on green packaging, establish a base of development strategy for lightweight product packaging, interview experts in industrial circles, introduce Fuzzy Logic in the research of the Grounded Theory, and analyse careful and objective factors of design and development of lightweight product packaging. Finally, this study conducts ANP on the general factors, and develops design and development rules of light-weight product packaging, by quantitative analysis and weight ranking. By the research findings, this study attempts to implement lightweight product packaging in the industry, and allow the consumers to have correct recognition of green lightweight packaging.

*Innovation in Design, Communication and Engineering – Meen, Prior & Lam (Eds)*
*© 2015 Taylor & Francis Group, London, ISBN 978-1-138-02752-7*

# Visual metaphor in poster design

Shih Hsi Yang
*College of Design, Chung Yuan Christian University, Taoyuan, Taiwan*

ABSTRACT: A rhetorical figure is a method that helps people express themselves more effectively. A metaphor is a figure of speech in which one thing is associated with another, where usually one is understood or described in terms of another. Previously, a metaphor was seen as a rhetorical and semiotic tool. However, it has now been applied to many fields, such as visual advertising, film and graphic design. This study focused on how visual metaphorical designs from different cultural backgrounds convey concepts in poster design. The method is to collect and organize the relevant metaphorical and classic posters from different cultures. This study concludes that a visual metaphor has a verbal advantage over rhetoric and semiotics, also because it can offer a powerful visual interpretation to the end users. The messages of the visual metaphor in posters are various. It is not necessarily serious, but can also be meaningful. Metaphorical images imply more than words can suggest. Being a designer, one needs to understand the cultural background well so that one could apply a metaphor in a design work well.

*Innovation in Design, Communication and Engineering – Meen, Prior & Lam (Eds)*
© 2015 Taylor & Francis Group, London, ISBN 978-1-138-02752-7

# Exploring service quality of social networking sites based on Kano model and innovation diffusion theory

Kuan-Yu Lin & Chih-Hua Chen
*Department of Information Networking and System Administration, Ling Tung University, Taiwan*

ABSTRACT: Web quality is a key factor affecting customer satisfaction and loyalty. Quality is determined chiefly by customer satisfaction. Only having uncovered and understood what customers really think about the service quality of social networking sites can the site service providers strengthen and design their sites based on the quality attributes that customers are concerned with, and to further attract them to continue to use.

This study thus applied Kano's two-dimensional model for quality for the analysis of customer requirements for quality of social networking sites. This study was also based on the Innovation Diffusion Theory, by whose characteristics factors were categorized to enable investigation of Kano's two-dimensional attributes of quality. By using said model, we hope to assist social networking site service providers in understanding what quality elements are currently most highly regarded and what elements need improvement, as well as how to satisfy customers by making changes to other services provided. Our conclusions and recommendations can serve as reference for the service providers and their managers in their operations.

*Innovation in Design, Communication and Engineering – Meen, Prior & Lam (Eds)*
© *2015 Taylor & Francis Group, London, ISBN 978-1-138-02752-7*

# Consumer-preferred pattern search approach for the design of Taiwan Tea package form

Chun-Wei Chen
*Ling Tung University, Taiwan, R.O.C.*

Shang-Te Tsai
*National Taiwan University of Science and Technology, Taipei, Taiwan, R.O.C.*

Sung-Chih Huang
*National Yunlin University of Science and Technology, Taiwan, R.O.C.*

ABSTRACT: The purpose of this study is to build a search method for the consumer-preferred design of the Taiwan Tea package form. Taiwan Tea is an important economical crop for Taiwan's economy. The tea package is a key factor for Taiwan Tea selling. To help create a better tea package, which is a more effective consumer-satisfying tea package, it is necessary to have a dedicated design for the consumer-preferred Taiwan Tea package form. In this paper, the architecture for comparing, searching analyzing and representing the consumer-preferred patterns of the Taiwan Tea package form were proposed. Through the approach, the designers can see the consumer-preferred design of Taiwan Tea package form. The proposed approach has been verified successfully on a design support system for the creation of comic figures. Experimental results show that, it not only appears promising in the domain of tea package design, but exhibits great commercial potential in the era of education.

*Innovation in Design, Communication and Engineering – Meen, Prior & Lam (Eds)*
*© 2015 Taylor & Francis Group, London, ISBN 978-1-138-02752-7*

# Developing a Kano-based evaluation model for innovation design

C.T. Wu & M.T. Wang
*Department of Industrial Design, National Kaohsiung Normal University, Kaohsiung, Taiwan*

C.S. Wu
*Department of Information Management, National Formosa University, Huwei, Taiwan*

ABSTRACT: This research focuses on developing a psychology-based evaluation procedure for innovative design. In this article, the extensive QFD developed by Wu has been adopted for the innovative design procedure. The major procedure of QFD is to identify the customers' needs for the product and then convert it into appropriate technical characteristics. According to the priorities of product characteristics, the prior engineering parameters will be identified to be the key requirements to redesign. The extension method will aid the Customers' Requirements (CRs) transform to product design attributes more comprehensively. For achieving the attractive design, we introduce the Kano model to construct the evaluation model. The proposed Kano-based evaluation procedure is mainly used in two stages of the innovative design. First, the evaluation process is used in QFD stage to help identify attractive customers' needs, the other is used in the extension stage to help assess concepts. The flowchart of the proposed innovative design procedure with psychology-based evaluation has also been developed. A case study, exercise equipment design, is adopted to explain and verify feasibility of the proposed approach.

*Keywords:* Kano model; extension method; matter-element; QFD

# Applying multimedia for combining harp music and photography design, with John Thomas' *Autumn* as an example

Yunn-Chu Chen
*Department of Visual Communication Design, Ling Tung University, Taiwan*

Wen C. Pai
*Department of Financial Engineering and Actuarial Mathematics, Soochow University, Taiwan*

ABSTRACT:   The mission of a musician is to attempt to transmit the precise thought of composers so that the audience members will understand and appreciate the work while they listen to the music. When performers interpret great music on their instruments, however, there is wide variance in the imaginations of listeners, according to their own experiences.

The purpose of this research is to design a complete image of music, based on the theory of harmony and colour psychology, by combining harp music and photography through multimedia, in order to interpret the music according to the musician's intent. Using reification of photography, obtained through the use of multimedia, composer John Thomas' *Autumn* will be played live to present the audience with the complete art of the composer.

*Innovation in Design, Communication and Engineering – Meen, Prior & Lam (Eds)*
*© 2015 Taylor & Francis Group, London, ISBN 978-1-138-02752-7*

# A study on the usability of a jeweller's workbench

Nien-Te Liu
*Department of Product Design, Shu-Te University, Kaohsiung, Taiwan*

Chang-Tzuoh Wu
*Department of Industrial Design, Kaohsiung Normal University, Kaohsiung, Taiwan*

Tsai-Ling Yang
*Department of Creative Product Design, Southern Taiwan University of Science and Technology, Tainan, Taiwan*

Hsin-Yi Tseng
*Graduate School of Applied Art and Design, Shu-Te University, Kaohsiung, Taiwan*

ABSTRACT:  Due to the special nature of the job, a long time working of jewellers on the workbench has led to the higher usability and demand of these professionals. Starting from such a concept, the study expects to find out usability issues of the jeweller's workbench through users and to be provided as a reference to related studies and the jeweller's workbench design of future designers. The study is driven by the latest information provided through interviews with experts and a questionnaire survey, with images of existing jeweller's workbenches firstly. To understand the usability issues of existing jeweller's workbench. According to the interviews conducted with experts, the existing jeweller's workbench have failed to meet expectations in terms of comfort level and efficiency, in which case the usability issue needs to be addressed mostly in terms of efficiency of use and subjective satisfaction. A questionnaire survey on usability also indicated similar issues of jeweller's workbench in terms of two usability indexes—efficiency of use and subjective satisfaction. Lastly, two outcomes were derived from the study; one, the key suggestions for the jeweller's workbench design and sequence of various design highlights by importance, the other, suggestions for the measurement guidelines of the jeweller's workbench design including, the height of the jeweller's workbench that users are used to—which is 80–90 cm, length and width they are used to are 99.7 cm and 63.1 cm respectively: a length of 15~30 cm, a width of 8~30 cm and a thickness of 4~8 cm is recommended for the large file bridge, a 10~15 cm length, 5~8 cm width and 3~4 cm thickness for the medium file bridge and 1~10 cm length, 1~5 cm width and 1~3 cm thickness for the small file bridge.

*Innovation in Design, Communication and Engineering – Meen, Prior & Lam (Eds)*
© 2015 Taylor & Francis Group, London, ISBN 978-1-138-02752-7

# An innovative light electric vehicle for caring shipping on logistic Transnet

Ming-tang Wang & Chien-Yu Liu
*Department of Industrial Design, National Kaohsiung Normal University, Kaohsiung, Taiwan, R.O.C.*

ABSTRACT: Online shopping has become popular. Now consumers can easily order goods from the Internet. When they pass through the Transnet logistics and distribution system, all goods can be delivered to consumers. The Internet is an innovative vehicle for delivering goods on Transnet which can also take care of the delivery. The process of iNPD (integrated New Product Development) has been operated to create a preliminary light electric vehicles' concept, which can be mounted on a delivery truck. It is convenient for delivering the goods even in narrow roads where the truck cannot pass. It can also be battery-operated and charged from the trucks to prevent insufficient electric endurance.

*Keywords*: Transnet electric vehicles caring design iNPD foldable

*Innovation in Design, Communication and Engineering – Meen, Prior & Lam (Eds)*
*© 2015 Taylor & Francis Group, London, ISBN 978-1-138-02752-7*

# Product design using Kansei-based creative associative thinking

Kun-Chieh Wang
*Department of Computer-Aided Industrial Design, Overseas Chinese University, Taichung, Taiwan*

Meng-Wei Ling
*Institute of Technological Product Design, Ling Tung University, Taichung, Taiwan*

ABSTRACT: To deal with the relationship between product images and design elements, a widely popular scheme, called "Kansei Engineering," was proposed by Nagamachi in 1989. However, among vast studies about product design using Kansei engineering, none concerned the most important issue in product design: creativity. In the first place, this study proposes a new Kansei manipulation procedure in which it combines the loop-type associative thinking process with the fuzzy Kansei engineering to design a new product matching current or future customers' requirements. A table horn is selected as the demonstration target. In total, four final Kansei images are obtained via market survey and statistical analysis. Seven design elements were obtained via product decomposition. Then the 7-scale semantic differential scheme and fuzzification scheme are used to quantify the qualitative properties of product image and design element respectively. A Kansei evaluation to 19 samples from 30 subjects is performed. Based on these evaluation data, the multi-linear regression and back-propagation neural network scheme are adopted to build the relationship between Kansei words and design elements. Then the above Kansei outcomes are used as the design basis and a loop-type creative associated thinking procedure, which includes six steps, is applied to develop a new product. In the end, a verification test is performed and a satisfying result is obtained, in which it reveals the new product gains at least 11.7% increase in customer satisfaction.

*Innovation in Design, Communication and Engineering – Meen, Prior & Lam (Eds)*
*© 2015 Taylor & Francis Group, London, ISBN 978-1-138-02752-7*

# Study on attractive factors of electronic products by environmental materials: Using notebook computers as example

Tung-Long Lin
*Department of Industrial Design, National Kaohsiung Normal University, Kaohsiung, Taiwan*

ABSTRACT: In design of consumer electronic products, there is application of a great amount of innovative and various materials. Usage and selection of different materials influence consumers' emotions at different dimensions. Nowadays, environmental consciousness is on the rise and environmental goods have gradually become the mainstream trend for future consumer markets.

Using the EGM interview of Emotional Engineering, this study conducted qualitative in-depth interviews and constructed an EGM evaluation structure of all subjects' constructs. Moreover, it used a 7-point scale to examine the consumers' identification with different materials applied to attractive features of notebook computers.

According to research findings, the means of attractive features of the samples are identified to some degree. Notebook computers of environmental materials are special, characteristic and of high-quality, the different materials vary slightly. According to the results of the features' analysis, there are 18 attractive features. With different materials, those with an artistic design background are more plentiful than those of a non-artistic design background. One-way ANOVA found that subjects of different backgrounds have positive comments on all results.

The research findings can help designers in recognizing the attractive features for consumers' and preference of notebook computers with different materials, thus facilitating the product design to better match the consumers' needs of perceptual attraction.

*Innovation in Design, Communication and Engineering – Meen, Prior & Lam (Eds)*
*© 2015 Taylor & Francis Group, London, ISBN 978-1-138-02752-7*

# A primary exploration of assistive devices' design by using QFD—the electric stander for CP children as example

I-Jen Sung

*Department of Innovation Design Engineering, National Kaohsiung First University of Science and Technology, Kaohsiung, Taiwan*

ABSTRACT: This study chose the standing designed for the children with cerebral palsy as the target, and invited 21 experts in the field of rehabilitation at Tainan City Assistive Device Resource Center and Chimei Hospital, as well as eight researchers to participate in this research. Data was collected from observation and interviews. Then, the Matrix of House of Quality of QFD was used to analyse, compare, and integrate the demands from medical institutions, rehabilitation centers, production units and users. The design guidelines of the standing frame are as follows: (1) Operability: the related accessories of the standing frame need to be simple and convenient for operation and adjustment with high intuitiveness; (2) Stability: when using the electric pushrod for structural design, attention needs to be paid to the tolerances of the fit between the base size and associated movable parts; (3) Economy: the compatibility of the parts needs to be considered when improving the design to enhance the operability and stability of the standing frame, in order to lower the production cost.

Then bring in the industrial design method and coordinate the numerous interacting factors and requirements to develop a standing frame of a brand new concept the rigorous development conditions. The standing frame has the following characteristics: (1) horizontal lifting system: with the electric pushrod equipped with a parallel four-bar linkage and a X-type lifting structure, the standing frame under the horizontal status can be adjusted upwards and downwards by 20 cm to match up with a child patient's figure, enabling a child patient to get on and off the standing frame conveniently and easily, reducing the base packaging material and lowering the transportation cost; (2) modularized frame: with the modularized frame design characteristic of the sharing of more than 80% of the parts, it is scalable from a small size to large sizes according to the increase of a child's height; moreover, three kinds of standing posture positions of "lean back", "stand upright" and "bend over" can be covered with a single frame equipped with different back plates; eventually, a total of five styles of standing frame products can be developed—namely, the "stand upright-lean back" style, "stand upright-bend over" style, large-size, medium-size and small-size, thus able to reduce the number of replacing and purchasing the standing frame and significantly decreasing approximately 70% of the economic burden; (3) electric tilt angle adjustment: with the electric pushrod as the power source of the angle adjustment and equipped with an angle gauge, the standing frame can be adjusted from the upright status of infinite section angle to the horizontal status. Its effort-saving operation is able to reduce the workload of a rehabilitation therapist; and the angle adjustment can enhance the rehabilitation effectiveness with the practical actions.

It is expected that the research results may help to build a development model for assistive devices and enhance the R&D standard of standers in Taiwan.

*Innovation in Design, Communication and Engineering – Meen, Prior & Lam (Eds)*
© *2015 Taylor & Francis Group, London, ISBN 978-1-138-02752-7*

# Analyzing needs and emotional preference factors of mobility aids for seniors

Jui-Che Tu, Kang-Chi Lin & Ying-Ju Ho
*Graduate School of Design, National Yunlin University of Science and Technology, Yunlin, Taiwan*

Chuan-Ying Hsu
*Department of Business Administration, Dayeh University, Changhua, Taiwan*

ABSTRACT: Everyone will face the aging problem. The decline of mobility is the main factor in the start of accelerated aging, because it will damage the quality of life, reduce social interaction, and then let the seniors negate the value of life. They get old faster and faster in a vicious circle, so there must be some ways to prevent their mobility decline. There are many mobility aids on the market to help the seniors deal with mobility problems, but they generally tag those items as "aging", and thus have a great sense of exclusion. Therefore, the designs of mobility aids for seniors need not only the considerations of function, but also the solutions for, in emotional factors perspective, reducing the negative perceptions of mobility aids. Then the using willingness and efficiency of mobility aids will be improved, and the seniors living in different lifestyles will have more robust mobility lives. It will mainly investigate the important literature related to Kansei Engineering and the seniors' mobility aids, and it will also conduct research on the lifestyle trends of the pre-seniors and the seniors. In order to establish the development foundation for matching the seniors' emotional-factor preferences and their product semantics, it will examine the research of lifestyle by A.I.O. inventory, and analyze the trends and the current situations of the seniors' lifestyles by using factor analysis and cluster analysis methods. Meanwhile, through Questionnaire Survey and Observation and Interviewing, it analyzes the seniors' consumer behaviors, and through the Analysis of Variation to understand the cognitions, the consumer attitudes, and the consumer demands to the mobility aids among the different groups of seniors. On the other hand, it will conduct surveys and cross-comparison in what are needed for the related industries to develop the seniors' mobility aids. And then it will obtain the emotional factors by analyzing and summarizing the seniors' mobility aids.

# Discussion on industrial design of cultural creative industrial pattern

Liang Chen
*Zhi Cheng College, Fuzhou University, Fuzhou, Fujian, P.R. China*

ABSTRACT: Industrial designers, as perfect conjunctions of technology, culture and art are the critical component of cultural creative industry as well as one of the core areas in transformation from creativities to productive forces. Industrial design of cultural creative industry has not only promoted self-innovation capability of enterprises but also be the strategic tools built by core competitiveness, therefore, governments have pay more and more attention to development of industrial design. Regarding to current status on industrial design of cultural creative industrial pattern mainly, this article makes discussions on developmental strategy of industrial design on cultural creative industrial pattern.

*Innovation in Design, Communication and Engineering – Meen, Prior & Lam (Eds)*
*© 2015 Taylor & Francis Group, London, ISBN 978-1-138-02752-7*

# Study on the interactive design of brand user experience for digital media environment

Qi Li & Yangduo Liao

*College of Physics and Information Engineering, Fuzhou University, Fuzhou, Fujian, P.R. China*

ABSTRACT: Taking many excellent works of interactive design at home and abroad as examples, this article chooses the user experience as the study object, combines with some basic theories and takes some relevant literature as reference to comparatively analyze the previous successful interactive design cases. Combining with theories and practices, the author summarizes the functions and the design features of the interactive design in brand advertising.

*Innovation in Design, Communication and Engineering – Meen, Prior & Lam (Eds)*
*© 2015 Taylor & Francis Group, London, ISBN 978-1-138-02752-7*

# Mechanical efficiency of a planetary-gear-train type automatic transmission

Artde Donald Kin-Tak Lam
*Xiamen Academy of Arts and Design, Fuzhou University, Xiamen, Fujian, P.R. China*

ABSTRACT: Planetary gear trains have the advantage of compactness, light weight, high speed reduction and capability of differential drive, so they are often used in transmission systems. The purpose of this paper is to propose a systematic methodology for the analysis of the mechanical efficiency of a planetary-gear-train type automatic transmission. Based on the Graph theory, a canonical graph representation of the kinematic structure of planetary-gear-train type automatic transmission is presented. Then, the equations of the mechanical efficiency of the planetary-gear-train type automatic transmission have been constructed. The design of an automatic transmission for the automobile is used as an example to illustrate the proposed methodology. The results of the paper provide a powerful tool for the design of the planetary-gear-train type automatic transmissions.

*Innovation in Design, Communication and Engineering – Meen, Prior & Lam (Eds)*
*© 2015 Taylor & Francis Group, London, ISBN 978-1-138-02752-7*

# Generate artistic fractal patterns based on complex system and Julia set

Artde Donald Kin-Tak Lam
*Xiamen Academy of Arts and Design, Fuzhou University, Xiamen, Fujian, P.R. China*

ABSTRACT: Under comparison of the fractal pattern and the traditional artistic pattern, the fractal pattern is a kind of the arts with mathematical characteristics. Based on complex function system and Julia set, the generation of artistic fractal patterns has been studied in this paper. First of all, the relationship between the fractal pattern and mathematical morphology parameters has been discussed; escape time algorithm was used to generate artistic fractal patterns. An actual fractal pattern generation model was established in order to discuss the control factors. Thus, a computer program has been developed to generate artistic fractal patterns. Finally, the patterns were generated by the developed program which could be applied to the artistic pattern design.

*Innovation in Design, Communication and Engineering – Meen, Prior & Lam (Eds)*
*© 2015 Taylor & Francis Group, London, ISBN 978-1-138-02752-7*

# Fractal characteristics of quilt patterns

Xiao-Yu Huang, Jun He & Artde Donald Kin-Tak Lam
*Xiamen Academy of Arts and Design, Fuzhou University, Xiamen, Fujian, P.R. China*

ABSTRACT: The pattern design is widely applied in daily life, offering unbounded pleasure to human life experience. The design evolution of the textile surface pattern is influenced by the cultural background, styles and design techniques of different periods. Based on fractal geometry, the main purpose of this article is to analyze the texture characteristics of the quilt pattern. First, a brief description of the fractal geometry is given and then, the characteristics of the surface shape of textile in quilt patterns are analyzed. Box-counting method is used to calculate the fractal characteristics of the textile of quilt pattern. Then, a computing program of the fractal characteristics is developed with the help of MATLAB. Through this program, the surface shape of textile in quilt patterns is described. The results of this article include a mathematical model to describe the observation of fractal characteristics of quilt patterns.

*Digital content & digital technology*

# Interactivity: Media convergence and digital media

Wenyan Xu

*Xiamen Academy of Arts and Design, Fuzhou University, Xiamen, Fujian, P.R. China*

ABSTRACT:   Convergence and digitization are closely associated with interactivity, which will be discussed in detail. This essay begins with the literature review of interactivity and digital media, which will lead to the ongoing debate of the strengths and weaknesses of interactivity in the new media era, say, the positive perspectives and negative perspectives of interactivity in the board range of digital media forms, in a sense that the development of digital media technologies has not only brought us Internet (both PC and Mobile) as brand new media, but also changed the way traditional media present themselves. Provided with the debate of whether or not interactivity is a positive perspective of digital media, this essay is going to focus on the question 'Is interactivity a defining feature of digital media?' It will be argued that, in many cases, if not most, interactivity is a defining feature of new digital media, but it is hard and maybe impossible for interactivity to become a defining feature in all sorts of digital media forms, for example, digital cinema productions.

*Innovation in Design, Communication and Engineering – Meen, Prior & Lam (Eds)*
*© 2015 Taylor & Francis Group, London, ISBN 978-1-138-02752-7*

# The ambition of the consumer on the mobile game: Tower of Saviors

Chih Cheng Sun & Sin Fatt Chun
*Department of Multimedia and Entertainment Science, Southern Taiwan University of Science and Technology, Nantai, Tainan, Taiwan, R.O.C.*

ABSTRACT: The advancement of the mobile application technology made mobile phones more than just a communication tool. Mobile phone functions also provide entertainment and necessary lifestyle applications. Furthermore, the growth of the mobile games market attracts game developers to participate in the development of mobile games. This study discusses the game experience and purchase intention of consumers on Tower of Saviors, a popular mobile phone puzzle battle card game. As a result of the market analysis and exploration of literature, a survey of mobile users who play the Tower of Saviors was conducted. The results of the survey show that: 1. Around 30 to 60 percent of the participants have a positive view of this game, but their purchase intention is not high. The reason for this is their status as students with limited purchasing ability. There were also participants with negative game experience; therefore, their purchase intention is easily affected. 2. Participants with less contact to the game, those with less gaming frequency, and those who do not often purchase mobile games are easily influenced in their game experience and purchase intention. Therefore, some recommendations are given based on these results for game developers and future relevant studies.

*Keywords*: mobile game; puzzle battle card game; purchase intention; game experience

*Innovation in Design, Communication and Engineering – Meen, Prior & Lam (Eds)*
*© 2015 Taylor & Francis Group, London, ISBN 978-1-138-02752-7*

# The impact of applying interactive multimedia materials to English teaching practices on children's learning of English—the case of K.C. English

Chih-Cheng Sun & Yu-Hao Lo
*Department of Multimedia and Entertainment Science, Southern Taiwan University of Science and Technology, Nantai, Tainan, Taiwan, R.O.C.*

Guan-Cheng Wang
*K.C. English, Kaohsiung, Taiwan, R.O.C.*

ABSTRACT: In recent years, as multimedia interactive technologies are widely applied in education, the teacher's role is also transformed from being an "initiator" to being a "mentor". Using multimedia animation and games as learning media enhances the children's learning interest and efficiency. It is not only restricted to a traditional fixed curriculum schedule but time can also be allocated in order to strengthen learning because practice can be repeatedly carried on parts that are not understood. These are the advantages of using interactive multimedia materials. This study investigates the effects of teaching English to children using interactive multimedia materials. Through the collection and analysis of relevant literature, the differences between traditional teaching and digital teaching were explored. Also investigated were the feasibility of implementing the application of interactive multimedia materials in teaching English to children, the impact on students of introducing game concept into such instructional materials, and the design principles of these materials. Through interview and observation, the methods and status of an after school learning center that uses its self-developed interactive multimedia materials in teaching was actually understood. This study found out that application of interactive multimedia materials can improve children's concentration during English lessons, increase interaction with teachers, develop the students' ability to express themselves, and enhance the students' learning motivation and effect.

*Keywords*: e-learning; interactive games; interactive multimedia materials; multimedia

*Innovation in Design, Communication and Engineering – Meen, Prior & Lam (Eds)*
*© 2015 Taylor & Francis Group, London, ISBN 978-1-138-02752-7*

# A study on the effects of subliminal stimulus messages in application of the image and sound mask effect mixing method

Jun He
*Xiamen Academy of Arts and Design, Fuzhou University, Xiamen, Fujian, P.R. China*

Hwang-Rae Kim & Jae-Woong Kim
*Department of Computer Science and Engineering, Kongju National University, Chungnam, South Korea*

ABSTRACT: This study aims to analyze the effects of subliminal stimulus messages delivered by humans unconsciously after certain sound or image clips are played simultaneously at every 25 frames or 50 frames or alternatively at certain intervals in application of the Mask Effect Mixing method. As a result, the frame rates of images decreased from 1/50 to 1/25, and as the play of images and sounds turns from asynchronism to synchronism, the rate of human memory enhanced. As time passed, the rate of memory gradually decreased. In the future, it is thought that the findings of this study will contribute to sales, behaviors, and communication effects as well as learning efficiency when applied to advertisement or learning.

# A communicational application to enhance the design process: Concept of Web-Based Communication Service for interior design

Xiao-Yu Huang & Kuo-Hsun Wen
*Xiamen Academy of Arts and Design, Fuzhou University, Xiamen, Fujian, P.R. China*

ABSTRACT:   This article stemmed from practice experience of being a professional interior designer. A number of questions arose mainly concerning the role of the interior designer in relation to the client. As the relationship between client and designer is always a dynamic one that changes over the course of the design process, this article focuses primarily on implementing communication modes to enhance the design process and achieve the design goals. As Internet has been broadly employed in our daily lives and the advancing of information communication technology, inevitably, communicational applications via Internet are counted as a useful tool to replace traditional communication ways conducted in design processes. By examining existing difficulties of communications and potentialities for the future design process, this article proposes a conceptual prototype of Web-Based Communication Service based on a user-centered design to enhance communication for the practice of interior design.

*Innovation in Design, Communication and Engineering – Meen, Prior & Lam (Eds)*
*© 2015 Taylor & Francis Group, London, ISBN 978-1-138-02752-7*

# Using fractal dimension to analyse the plane formation for the lacquer painting

Guo-Hua She
*Xiamen Academy of Arts and Design, Fuzhou University, Xiamen, Fujian, P.R. China*

ABSTRACT: Lacquer painting is integrated with the lacquering integration of the lacquer arts in painting a brand-new species. The samples discussed in this article focus on lacquer painting, using the fractal dimension to probe into the complexity and similarity in characteristics of the lacquer painting, to serve as a basis of the plane formation for the lacquer painting. The results of the study show that through the use of a mathematical model to represent the artistic style of lacquer painting, and the use of scientific quantitative analysis, the information of the plane formation for the lacquer painting can be derived.

*Innovation in Design, Communication and Engineering – Meen, Prior & Lam (Eds)*
*© 2015 Taylor & Francis Group, London, ISBN 978-1-138-02752-7*

# Automatic segmentation of motion capture data of Quan-zhou chest-clapping dance

Si-Xi Chen, Jian-wei Li & Jian-de Lin
*College of Physics and Information Engineering, Fuzhou University, Fuzhou, China*

ABSTRACT:   Quan-zhou chest-clapping dance is one of the national intangible cultural heritages of China, and the motion capture technique is an important method to realize its protection and inheritance. In this paper, we present the automatic segmentation technique of the motion capture data, in which the motion data is automatically segmented through KPCA combined with PPCA to reduce the dimension and project the 56 dimensional data in 2 dimensional space; formulate the feature function to obtain the derivative of projection error, detect the segmentation point of data through analyzing the change of geometric features to realize the automatic segmentation. It is indicated in the experiment that the motion capture technique has certain feasibility.

*Keywords*:   Quan-zhou chest-clapping dance; automatic segmentation; dimensionality reduction; motion capture

*Electrical & electronic engineering*

*Innovation in Design, Communication and Engineering – Meen, Prior & Lam (Eds)*
*© 2015 Taylor & Francis Group, London, ISBN 978-1-138-02752-7*

# Design and implementation of a multi-axis motion controller using SoPC technology

Ying-Shieh Kung, Yu-Jen Chen & Shi-Cun Liu
*Department of Electrical Engineering, Southern Taiwan University of Science and Technology, Tainan, Taiwan*

Hsin-Hung Chou
*Mechanical and System Research Laboratories, Industrial Technology Research Institute, Hsinchu, Taiwan*

ABSTRACT:  Development of a multi-axis motion control system with a compact size has become a popular trend in the world. Therefore, the goal of this paper is to implement a multi-axis servo controller and the motion trajectory planning within one chip. At first, SoPC (System on a Programmable Chip) technology, which is composed of an Altera FPGA (Field Programmable Gate Arrays) chip and an embedded soft-core Nios II processor, is taken to develop the multi-axis motion controller. The proposed multi-axis motion control IC has two modules. The first module is an embedded soft-core Nios II processor which is used to realize the motion trajectory planning by software which includes step position response, circular motion trajectory, helicoid motion trajectory, etc. The second module is presented to realize the multi-PMSM's position/speed/current controllers by hardware. And VHDL (VHSIC Hardware Description Language) is applied to describe the overall multi-axis controller behavior. Therefore, a fully digital multi-axis motion controller will be implemented by a single FPGA chip. Because FPGA have the advantages of fast computation, parallel processing and software-hardware co-design, it will apparently improve the performance of the multi-axis motion control system. Finally, to verify the effectiveness and correctness of the proposed multi-axis motion control IP, a three-axis motion platform (XYZ table) is constructed and some experimental results are presented.

*Innovation in Design, Communication and Engineering – Meen, Prior & Lam (Eds)*
© *2015 Taylor & Francis Group, London, ISBN 978-1-138-02752-7*

# Novel non-direct signal detect probing and modeling developed by the Biot-Savart theorem

Sung-Mao Wu, Hsin-Fang Li, Yin-Hsiu Yeh & Ren-Fang Hsu
*Micro Electrical Packaging Laboratory, Department of Electrical Engineering, National University of Kaohsiung, Taiwan, R.O.C.*

Cheng-Fu Yang
*Department of Chemical and Materials Engineering, National University of Kaohsiung, Taiwan, R.O.C.*

ABSTRACT:   In this study, the main keypoint is non-contacting measurement. The traditional contacting probe has numerous disadvantages, including low mobility, high cost, high damaged rate, and so on. A non-contacting probe designed efficacy with broadband, well reproducibility, well directivity and low loss coupling to transmit or receive a signal between different chips or ICs in 3DIC application, and establishing physical model up to 10 GHz with coupling effect and loop effect to restore the signal radiation. In addition deriving from the electromagnetic theorem discovers either the current flow or the scattering parameter. A complete reduction theory to compensate the coupled signal measuring by the non-contacting probe is needed. The result of the restore theory is highly matched to the model with lumped elements, and the result of the derived theory is similar to actual measurement that can detect the signal in ICs.

*Innovation in Design, Communication and Engineering – Meen, Prior & Lam (Eds)*
© *2015 Taylor & Francis Group, London, ISBN 978-1-138-02752-7*

# Sensorless sliding-mode control of Permanent Magnet Synchronous Motor based on high frequency injection method

Wang Ming Syhan, Ika Noer Syamsiana & Taryudi
*Department of Electrical Engineering, Southern Taiwan University of Science and Technology, Tainan, Taiwan*

ABSTRACT: This paper proposes a sensorless sliding mode control to achieve a good performance of the Permanent Magnet Synchronous Motor (PMSM) at any speed. High frequency signal is injected in α-axis winding in order to detect rotor position. Sliding-Mode Observer (SMO) based on stationary reference frame currents and back Electromagnetic Force (EMF) are proposed to extract back EMF to estimate the rotor position. High performance proposed sensorless control drive is simulated by Matlab/Simulink. The simulation results show effectiveness of the proposed system at any speed.

*Innovation in Design, Communication and Engineering – Meen, Prior & Lam (Eds)*
*© 2015 Taylor & Francis Group, London, ISBN 978-1-138-02752-7*

# The fast algorithm of short-term unit commitment

Tai-Ken Lu, Hao-Hsuan Hsu & Kai-Wen Huang
*Department of Electrical Engineering, National Taiwan Ocean University, Keelung, Taiwan*

ABSTRACT:   The power industry is one of the major energy industries, and how to dispatch the generators to minimize the operation cost with system security is the most important target for power dispatchers. When power dispatchers propose a day-ahead generation plan, they usually consider load balanced, unit characteristics, fuel consumption and other system constraints, that are also considered in the short-term unit commitment problem. This paper proposes the short-term unit commitment model with mixed integer linear programming that the objective function is total generation cost with unit constraints and system constraints. Due to the computing time of solving the unit commitment problem with a complete model is very long; it cannot support a proposal to the power dispatchers in a short time. So this paper also proposes the simplified model combined with a two-stage algorithm and using the simplified model's solution as the parameter of the complete model to speed up the solution. And this paper uses Taipower's nine thermal units to verify accuracy.

*Innovation in Design, Communication and Engineering – Meen, Prior & Lam (Eds)*
*© 2015 Taylor & Francis Group, London, ISBN 978-1-138-02752-7*

# IoT real-world monitoring the household electric power

Jih-Fu Tu
*Department of Electronic Engineering, St. John's University, Taiwan*

Chih-yung Chen
*Department of Information Management, St. John's University, Taiwan*

Chien-Min Ou
*Department of Electronic Engineering, Chien Hsin University of Science and Technology, Taiwan*

ABSTRACT:    To accomplish the purpose of saving energy, we proposed an AMI, which is based on an embedded system platform as the control center of the household power monitor, to achieve the smart-meter with control of a micro-processing chip, but will analyze, calculate, compute, and communicate functions to the household electrical-appliances. For transform and display of the household-power information, we respectively exploit RF technology of Zigbee and Visual Basic.NET to create the household-power server with HMI format and IEC 62056-53 device language message specification. We also measure the household-power information to a power-heater with the proposed AMI equipment and the HIOKI 3286-20 power meter. The maximum deviation of power, voltage, and current are 1 watter, 1.83 V, and 0.04 A, individually.

*Keywords:*   smart meter; advanced metering infrastructure; and Zigbee

*Innovation in Design, Communication and Engineering – Meen, Prior & Lam (Eds)*
© 2015 Taylor & Francis Group, London, ISBN 978-1-138-02752-7

# A novel Hybrid Particle Swarm Optimization for optimal power dispatch problem

C.H. Huang, C.C. Huang, Y.W. Chen & Y.C. Liang
*Department of Automation and Control Engineering, Far East University, Tainan City, Taiwan*

H.C. Lee & H.J. Chiu
*Department of Mechanical Engineering, Far East University, Tainan City, Taiwan*

ABSTRACT:   This paper presents a Hybrid Particle Swarm Optimization (HPSO) for the solution of the optimal power flow based on Current-Power Optimal Power Flow (CPOPF) with both continuous and discrete control variables that was called Mixed-Integer Optimal Power Flow (MIOPF). The continuous control variables modeled are unit active power outputs and generator-bus voltage magnitudes, while the discrete ones are transformer-tap settings and shunt capacitor devices. The feasibility of the proposed method is exhibited for standard IEEE 30 bus system, and it is compared with other stochastic methods in terms of solution quality, convergence property, and computation efficiency.

*Innovation in Design, Communication and Engineering – Meen, Prior & Lam (Eds)*
*© 2015 Taylor & Francis Group, London, ISBN 978-1-138-02752-7*

# Design and implementation of an advanced LED headlight with Adaptive Front-lighting System

Chih-Lung Shen, Wang-Ta Sung, Hung-Tse Kung & Tsair-Chun Liang
*Department of Electronic Engineering, National Kaohsiung First University of Science and Technology, Yanchao District, Kaohsiung, Taiwan, R.O.C.*

ABSTRACT: This paper proposes an Electric Driver (ED) for LED headlight with Adaptive Front-lighting System (AFS) of vehicle. The ER mainly contains two parts. One is power stage and the other is control unit. The power stage is a single-stage structure, instead of multi stages, such that it can achieve high efficiency, low cost, and compact features. Even though the proposed ER is a single-stage, all LED strings can be controled individually and then, easily accomplish AFS functions. The control unit is carried out by two Microchip MCUs, which controls LED on/off, performs active switch controlling, determines LED current, functions circuit protection, and fulfills dimming feature. To verify the ER, a prototype is built, simulated, and tested. From simulations and hardware measurements, the feasibility of the proposed ED is demonstrated. Furthermore, a series of experiments, such as environmental test, protection test, and burning examination, are dealt with. The practical results proof that the ER can meet the requirements for production.

*Innovation in Design, Communication and Engineering – Meen, Prior & Lam (Eds)*
*© 2015 Taylor & Francis Group, London, ISBN 978-1-138-02752-7*

# Study of magnetic properties on magnetic nanostructure array with different pore diameter on Anodic Alumina Oxide template

Chen-Hao Luo, Jenn-Kai Tsai, Teen-Hang Meen & Tian-Chiuan Wu
*Department of Electronic Engineering, National Formosa University, Yunlin, Taiwan*

ABSTRACT: In this study, porous alumina template and nanostructures were used as the defects of the magnetic thin film. By tuning the anodize processes, that is changing the anodize voltage, time of the electrolyte process, time of reaming process, it is possible to control its geometry. Samples were obtained by two-step anodize processes. We used 0.3 M oxalic acid solution as electrolyte, an anodize voltage of 20–60 V, and a temperature of 10°C. Films with antidot array have been prepared by the sputtering of $Ni_{80}Fe_{20}$ onto anodic alumina membrane templates. The pores diameter varies from 26 to 90 nm. A counterpart continuous thin film grown on Al substrate was also prepared. The coercivities of the antidot arrays are not only greater than those of un-patterned films, but it is changed by the films integrity and the pore diameter. The results show that the pore diameter becomes higher, the coercivity increases. It is presumed that the parameters of the size and type of defects hinder the movement of the magnetic moment.

*Innovation in Design, Communication and Engineering – Meen, Prior & Lam (Eds)*
*© 2015 Taylor & Francis Group, London, ISBN 978-1-138-02752-7*

# Design and implementation of an Extra-High Step-Up Converter for PV power system

Chih-Lung Shen, You-Shen Shen & Tsair-Chun Liang

*Department of Electronic Engineering, National Kaohsiung First University of Science and Technology, Yanchao District, Kaohsiung, Taiwan, R.O.C.*

ABSTRACT: A novel extra-high step-up converter is proposed in this paper, which can boost a low input voltage to an extra-high level. Accordingly, it is much more suitable for PV or fuel-cell power system. The proposed structure incorporates two coupled inductors and switch capacitors so as to stack up the voltage for high output voltage accomplishment. The proposed step-up converter essentially possesses the outstanding features, such as continued input current, single-switch configuration, galvanic isolation, lower turns ration, high voltage gain, and lower component count. As compared to other types of step-up converter, under the same duty ratio and turns ration, the proposed converter can achieve higher voltage gain than conventional ones. That is, by means of increasing turns ration, the duty ratio of the converter can be reduced significantly for stable operation. Since the energy stored in leakage inductance can be recycled totally, the efficiency can be improved. In the paper, a prototype is carried out for simulation, test, and practical measurements. The simulated results and measured waveforms have verified the feasibility and validation.

*Engineering materials & industrial applications*

*Innovation in Design, Communication and Engineering – Meen, Prior & Lam (Eds)*
*© 2015 Taylor & Francis Group, London, ISBN 978-1-138-02752-7*

# Golf accessory designs for young female adults with Kansei Form Composition approach

Hui-Fen Pan
*Graduate School of Applied Arts and Design, Shu-Te University, Yanchao, Kaohsiung, Taiwan*

Chyun-Chau Lin
*Department of Product Design, Shu-Te University, Yanchao, Kaohsiung, Taiwan*

ABSTRACT: In the highly diversified market, effectively fulfilling emotional or kansei needs of target users is a crucial issue. This article employs Kansei Form Composition (KFC) approach to develop styling designs of golf accessories for young female adults. The KFC approach not only integrates kansei thinking and morphological analysis but also suggests factor weights and flash cards in the processes. During the analyzing stages, nine related kansei adjectives and nine form elements are selected as main design factors. With the flash cards, the researchers can indicate the styling factors literally, visually and quantitatively. Heads of the driver and the 7-iron, and the caddie bag are developed based on the morphological compositions of related flash cards. Moreover, user evaluations by the young female adult golfers result in high agreement.

*Green technology & architecture engineering*

*Innovation in Design, Communication and Engineering – Meen, Prior & Lam (Eds)*
© *2015 Taylor & Francis Group, London, ISBN 978-1-138-02752-7*

# The study of BOP group's lifestyle and Sustainable Consumption Behavior design

Jui-Che Tu & Min-Chieh Shih
*Graduate School of Design, National Yunlin University of Science and Technology, Taiwan*

Chuan-Ying Hsu
*Department of Business Administration, Dayeh University, Taiwan*

ABSTRACT: Under global climate change and the situation of limited resources, it is urgent to develop a sustainable environment; moreover, every citizen on earth has a responsibility of facing the issue of sustainability. The researchers have set "3C product" as the theme for the target BOP group, conducted questionnaires to analyze data, and discussed the relations between BOP group's lifestyle, demographic variables as well as the concept of sustainable consumption and actual consumer behavior.

*Keywords*: Bottom of the Pyramid (BOP); lifestyle; Sustainable Consumption Behavior (SCB); Sustainable of Consumption & Production (SCP)

*Innovation in Design, Communication and Engineering – Meen, Prior & Lam (Eds)*
© 2015 Taylor & Francis Group, London, ISBN 978-1-138-02752-7

# Green management: A case study of HYA cable television's account management process improvement

Kuan-Ling Lai & Huann-Ming Chou
*Department of Mechanical Engineering, Kun Shan University, Tainan, Taiwan*

Ya-Ling Huang
*Department of Visual Communication Design, Kun Shan University, Tainan, Taiwan*

ABSTRACT: This study explores how to incorporate a model of green management to improve the account management procedures in digital cable television companies. Experimental zones can be set up to enhance overall business performances by improving corporate workflow, as well as lowering operating cost, manpower demand, and data transmission expenses. This study takes advantage of the interactive feature of digital cable television network transmission to increase the convenience and transparency in the communication between the service provider and its clients. Through the two-way communication interface of digital TV, the service provider can receive instant comments and suggestions from users to guide future improvements and make immediate remedies to meet users' expectations.

*Keywords*: Cable TV; green management; digitization

*Innovation in Design, Communication and Engineering – Meen, Prior & Lam (Eds)*
*© 2015 Taylor & Francis Group, London, ISBN 978-1-138-02752-7*

# Design of an advanced retractable multi-speaker headphone

Yu Tin Chao, Ya Lin Yu, Jia Yush Yen & Kang Li
*Department of Mechanical Engineering, National Taiwan University, Taipei, Taiwan, R.O.C.*

Yi Chien Lo
*Department of Optics and Photonics, National Central University, Taoyuan, Taiwan, R.O.C.*

You Ting Liao
*Department of Mechanical Engineering, National Taiwan University, Taipei, Taiwan, R.O.C.*

Lai Chung Lee
*Department of Interaction Design, National Taipei University of Technology, Taipei, Taiwan, R.O.C.*

ABSTRACT: This study sets up a complete mathematical model of a headphone. The whole model integrated circuit systems, mechanical systems and acoustic systems. By transforming all the systems to mechanical systems, we use the equivalent circuit method to derive the governing equation. From comparing it with different structure analysis, this study proposes an improved stucture design for better sensitivity and designing the stucture as a retractable headphone.

*Innovation in Design, Communication and Engineering – Meen, Prior & Lam (Eds)*
*© 2015 Taylor & Francis Group, London, ISBN 978-1-138-02752-7*

# Production of nano structured stainless steel surface through the Electrolytic Plasma Technology

Te-Ching Hsiao & Jhao-Jhong Su
*Department of Mechanical Engineering, National Kaohsiung University of Applied Science, Kaohsiung, R.O.C.*

Wei-Cheng Hsieh
*Department of Mechanical Engineering, Cheng-Shiu University, Kaohsiung, R.O.C.*

Cheng-Yi Chen
*Department of Electrical Engineering, Cheng-Shiu University, Kaohsiung, R.O.C.*

Chao-Ming Hsu
*Department of Mechanical Engineering, National Kaohsiung University of Applied Science, Kaohsiung, R.O.C.*

ABSTRACT: The aim of this research is to adjust the input voltage, impulse type and processing time to control plasma electrolytic technology. At the same time, to experiment the effect to the plasma electrolytic technology from factors such as base material surface roughness, processing time, and the frequency and duty ratio. For surface roughness, to optimize the experiment accuracy, the surfaces of all specimens used for this experiment have been first polished, then further grounded with #240, #400, #600 and #800 abrasive bands to produce a different surface roughness. Furthermore, the processing time is also the key parameter and is taken for 2~600 seconds. In impulse type, adjust different impulse work time to observe the variation of the frequency and the duty ratio. The work time adjusts from 15 μs to 1500 μs with the impulse rest time fixed at 5 μs. The voltage used for this study is 220 V~240 V. Having been processed, all specimens will be observed using a SEM (Scanning Electron Microscope), measured with an AFM (Atomic Force Microscope) including their contact angles.

Innovation in Design, Communication and Engineering – Meen, Prior & Lam (Eds)
© 2015 Taylor & Francis Group, London, ISBN 978-1-138-02752-7

# Process control parameters of nano structured stainless steel surface through the plasma electrolytic technology

Te-Ching Hsiao & Jhao-Jhong Su
*Department of Mechanical Engineering, National Kaohsiung University of Applied Science, Kaohsiung, Taiwan, R.O.C.*

You-Qing Cao & Ah-Der Lin
*Department of Mechanical Engineering, Cheng-Shiu University, Kaohsiung, Taiwan, R.O.C.*

Chao-Ming Hsu
*Department of Mechanical Engineering, National Kaohsiung University of Applied Science, Kaohsiung, Taiwan, R.O.C.*

ABSTRACT:   The aim of this research is to confirm the plasma electrolytic reaction zone for the reaction system assembled. At the same time, research shows how the plasma reacts when the clearance between two polarities is adjusted. The specimen used SUS316 stainless steel and the electrolytic solution was water-based sodium bicarbonate. All specimens are processed by polishing and grinding. To identify the zones of plasma electrolysis by measuring the voltage–current curve, the voltage increased continuously and observed the current variation at the same time. According to the voltage value is shown that the plasma electrolysis at 220 V~240 V is in a stable zone. Therefore, the voltage used is 220 V~240 V and the frequency was 50 kHz. In this research, the current density should be used as an index parameter for the plasma electrolytic reaction zone and the voltage value result is more accurate.

# A novel home-ventilation soundproof structure design based on nature-landscape model

Ya Lin Yu, Yu Tin Chao & Jia Yush Yen
*Department of Mechanical Engineering, National Taiwan University, Taipei, Taiwan, R.O.C.*

ABSTRACT:   In this article we bring up two soundproof structures based on muffler theorem which is composed of several channels parts and can be applied to interior design. Unlike general anti-noise structure design, the geometry shape design in this study is adopted by the landscape model. The property of this new design is that the most incident sound pressure will be blocked by the changes of cross section area within the structure region, but also provide enough region for ventilation. Simulations of the overall sound field demonstrate the structure can reduce the incident noise while allowing wind to pass through.

*Innovation in Design, Communication and Engineering – Meen, Prior & Lam (Eds)*
*© 2015 Taylor & Francis Group, London, ISBN 978-1-138-02752-7*

# Growth $Fe_2O_3$ films on stainless steel for solar absorbers

Sean Wu
*Department of Electronics Engineering and Computer Science, Tung Fang Design Institute, Hunei District, Kaohsiung City, Taiwan*

Kuan-Ting Liu
*Department of Electronic Engineering, Cheng Shiu University, Niaosong District, Kaohsiung City, Taiwan*

ABSTRACT: This study is to investigate the characteristics of high temperature solar selective absorbing films on the 304 stainless steel (304 SS) substrates. The $Fe_2O_3$ films were grown on specular SS substrates to be absorbers by high thermal process at 850 °C, 900 °C, 950 °C, 1000 °C and 1050 °C. The crystalline structure, surface microstructure and optic properties of the films were determined by X-Ray Diffraction (XRD), Scanning Electron Microscopy (SEM) and UV/visible spectroscopy (UV-VIS-NIR Spectrophotometer, 0.25–2.5 um). The $Fe_2O_3$ films prepared at 1000 °C of this structure displays a high absorptance of 0.92 and a low thermal emittance of 0.38 at 82 °C.

*Keywords:* 304 stainless steel; solar absorbers; absorptance; emittance

*Innovation in Design, Communication and Engineering – Meen, Prior & Lam (Eds)*
© *2015 Taylor & Francis Group, London, ISBN 978-1-138-02752-7*

# Study on the influence of electrolyte concentration on the Dye-Sensitized Solar Cells performance and long-term stability using Taguchi method

Jenn-Kai Tsai, Tian-Chiuan Wu, Jia-Song Zhou & Teen-Hang Meen
*Department of Electronic Engineering, National Formosa University, Yunlin, Taiwan*

ABSTRACT: Since electrolyte is a critical factor for high performance and long-term stability of Dye-Sensitized Solar Cells (DSSCs), finding an optimal composition of electrolyte is very important. Taguchi method has been used to design 16 types of gel electrolyte and is applied to DSSCs. The photo electrochemical characteristics and Electrochemical Impedance Spectroscopy (EIS) have been characterized. The S/N ratio is calculated from the efficiency of each group. Then the importance of each component in the electrolytes is measured. In this paper, we focus on the effect of the concentration of 1,2-dimethy l-3-propylimid-azolium iodide (DMPII) and 4-tertbutylpyridine (4-TBP) in 15 wt% gel electrolyte. The experiment result shows that the increase of DMPII reduces electrolytes diffusion impedance ($R_D$) and enhances short-circuit current density ($J_{SC}$). 4-TBP enhances the open-circuit voltage ($V_{OC}$) and both have an obvious effect on DSSCs performance and long-term stability.

*Green technology material*

*Innovation in Design, Communication and Engineering – Meen, Prior & Lam (Eds)*
© *2015 Taylor & Francis Group, London, ISBN 978-1-138-02752-7*

# Structure and dielectric properties of lower valence compensation for $Ba_{1-x}Na_x(FeTa)_{0.5}O_3$ by reaction sintering process

Yin-Lai Chai, Shuo-Chen Lin & Yee-Shin Chang
*Department of Jewelry Technology, Dahan Institute of Technology, Hualien, Taiwan*
*Department of Electronic Engineering, National Formosa University, Huwei, Yunlin, Taiwan*

ABSTRACT: Complex perovskite oxides of $Ba_{1-x}Na_x(FeTa)_{0.5}O_3$ (BNFT, x = 2, 4, 6, 8, 14, and 20 mole%) dielectric ceramics doped with lower valence compensation ions were prepared by a reaction-sintering process. The BNFT dielectrics were sintered at 1300°C for 8 h without calcinations involved. Due to low melting point of $Na_2O$, the sintering temperature of dielectric decreased. The density of sintered dielectrics increased with increasing $Na^+$ ion concentration and it cause BNFT dielectric surface glossy. The relative density of sintered dielectrics was higher than 89.5%. The dielectric constant of $Ba_{0.94}Na_{0.06}(FeTa)_{0.5}O_3$ 1300°C sintered ceramic was about $9 \times 10^5$, measured at room temperature at frequency of 1 kHz. The temperature-dependent loss curves measured in temperature range of $-15$~90°C and frequency of 1 k~1 MHz indicated that the samples exhibit a typical relaxor behavior with a broad temperature range which was contributed by orientation and space charge polarization.

*Keywords:* dielectric properties; reaction-sintering process; complex perovskite oxides; BFN

Innovation in Design, Communication and Engineering – Meen, Prior & Lam (Eds)
© 2015 Taylor & Francis Group, London, ISBN 978-1-138-02752-7

# A Low Dropout Regulator with low power consumption

Jih-Fu Tu
*Department of Electronic Engineering, St. John's University, Taiwan*

Chih-Yung Chen
*Department of Information Management, St. John's University, Taiwan*

ABSTRACT: The most novel electronic consumer products with digitalized high voltage are processed by a linear regulator with low voltage Bandgap circuit. In this paper, a 5 volts bandgap reference voltage is used to LDO to produce an output then to provide to bandgap and LV (lower voltage) circuit. We experiment the proposed Low Dropout Regulator, while output is 12 V, to find when the maximum load-current is 100 mA the gain is 60.4 dB and phase margin is 41°, and while the minimum load-current is 20 mA then the gain is 73.3 dB and the phase margin is 90°, individually.

*Keywords*: bandgap circuit; low dropout regulator; linear regulator

*Industrial design & design theory*

# Study on children's web page preferences through participatory design

Yu-Ming Chang

*Department of Multimedia and Entertainment Science, Southern Taiwan University of Science and Technology, Tainan, Taiwan*

I-Chen Wang & Min-Yuan Ma

*Department of Industrial Design, National Cheng Kung University, Tainan, Taiwan*

ABSTRACT:   This study introduced the process of participatory design to child-oriented web pages. Through sample planning, selection, and testing, the relationship between the form and configuration of web page interface factors and the preferences of the subjects were examined. The experimental results showed that the outcome from children's participatory design could efficiently improve their preference. In addition, the "form and position of the main menu", the "content of the central section", and the "display mode of the material", which are considered to be the three important interface factors affecting user preference, can be standardized. The best configuration according to preference matching was summarized in this study for interfaces with different intended uses. This is to offer some form of reference for child-oriented web page research and design in the future.

*Innovation in Design, Communication and Engineering – Meen, Prior & Lam (Eds)*
*© 2015 Taylor & Francis Group, London, ISBN 978-1-138-02752-7*

# Ejector nozzle design by numerical simulation for the ejector type air-conditioner system

Chang-Ren Chen, Han-Te Wu, Jung-Feng Su & De-Lu Chen
*Department of Mechanical Engineering, Kun Shan University, Tainan, Taiwan*

Atul Sharma
*Rajiv Gandhi Institute of Petroleum Technology, U.P., India*
*Kun Shan University, Tainan, Taiwan, R.O.C.*

ABSTRACT: This paper presents numerical analyses for the study of the ejector nozzle of a solar aided ejector-based air conditioner system which uses water as working fluid. Analyses of steady-state Computational Fluid Dynamics (CFD) were conducted in order to corroborate and optimize the results of previous one dimensional studies in a three dimensional simulation. Several valuable discoveries were found through the simulation: Optimal conditions of geometry for known mass flow rates and temperatures were found; different dimensions of the nozzle were found as optimal for the same conditions proposed in previous literature; it was also proven that geometries and initial conditions, proposed in other literature, don't always yield physically reasonable results.

*Keywords:* ejector; nozzle; design; solar; simulation; air conditioner

*Innovation in Design, Communication and Engineering – Meen, Prior & Lam (Eds)*
*© 2015 Taylor & Francis Group, London, ISBN 978-1-138-02752-7*

# Design of drink Vending Machines with computationally improved refunding system based on Petri Nets

Yen-Liang Pan & Wei-Hsiang Liao
*Department of Avionic Engineering, Airforce Academy, Koahsiung, Taiwan*

Cheng-Fu Yang
*Department of Chemical and Materials Engineering, National University of Kaohsiung, Kaohsiung, Taiwan*

ABSTRACT: This paper focuses on the use of Petri Nets (PNs) to model a drink Vending Machine (VM) system. In fact, a VM is a kind of machine that can provide various drinks, snacks, beverages, tickets, and other products for all consumers automatically. The greatest advantage of a VM is that it is able to sell products without a cashier at any time. In other words, a VM brings much convenience to our lives. However, it also causes a great deal of consuming dissension once the product or refunding system of one VM gets in trouble. Therefore, how to design one stable product and refunding system in one VM seems an important issue. In existing literature, time Petri net is used to design one coffee VM with excellent refunding system. In our study, however, it seems too complex to practice in a real world. In this paper, the authors try to develop a computationally improved refunding system using much simpler Petri nets. Besides, one real world VM is used as evidence to our model. According to the experimental results by using the famous software HPsim, it reveals the designed model is correct and practical. The most important, the proposed model is much simpler than existing literatures.

*Innovation in Design, Communication and Engineering – Meen, Prior & Lam (Eds)*
© *2015 Taylor & Francis Group, London, ISBN 978-1-138-02752-7*

# Application of Delphi method in constructing evaluation indicators of design students' core competencies

Shu-Ping Chiu
*Department of Digital Media Arts and Design, Fuzhou University, Xiamen, China*

Jui-Che Tu
*Graduate School of Design, National Yunlin University of Science and Technology, Yunlin, Taiwan*

Wei-Cheng Chu
*Department of Fashion Design, Shu-Te University, Kaohsiung, Taiwan*

Li-Wen Chuang
*Department of Digital Media Arts and Design, Fuzhou University, Xiamen, China*

ABSTRACT:   In recent years, the college enrollment rate has increased substantially but the unemployment rate of college graduates has also gone up. Subrahmanyam (2013) indicated that technical and vocational education can effectively solve the problem of unemployment and has become an international research issue presently. For adapting trends of globalization and knowledge economy as well as a rapid change of labor market and surrounding environment, this generation of young people should possess not only professional skills but also core employment ability that meets new economic requirements when entering the workforce. Morris Chang, Chairman of TSMC, says current severe industry problem in Taiwan is caused by gap between education and industry. For the sake of implementing development of pragmatic and practical curriculum in vocational and technical colleges as well as providing courses in conformity with industrial requirements and in favor of increase in TVET (Technical and Vocational Education and Training) students' career competitiveness, the Ministry of Education actively plans for and establishes indicators for various study groups' professional competences. In such context, this study aims at developing a set of evaluation indicators applicable to core competences of design students in Taiwan. It utilizes expert decision of Delphi method to establish evaluation indicators. Findings Core competencies including "professional capabilities" and "key skills". Professional competence refers to the ability required by the professional field; key competence refers to the common capability in a workplace. Professional competence includes five aspects: design literacy, design theory, operational practice, design application, and design planning. Key competence includes four aspects: attitude ethics, creative thinking, self-management, and analysis and application. Each aspect contains several assessment items, which all achieve expert consistence standard and expert consensus. Definite evaluation indicators developed here can be used as standards for related education units to assess students' competences, as well as bases for the industry to choose talents and make appraisal in the future. These are major contributions of this study.

*Innovation design & creative design*

*Innovation in Design, Communication and Engineering – Meen, Prior & Lam (Eds)*
*© 2015 Taylor & Francis Group, London, ISBN 978-1-138-02752-7*

# Study on mascot design strategy of brands of happiness

Jui-Che Tu
*Graduate School of Design, National Yunlin University of Science and Technology, Yunlin, Taiwan*

Ya-Wen Tu
*Department of Commercial Design, Chienkuo Technology University, Changhua County, Taiwan*
*Graduate School of Design, National Yunlin University of Science and Technology, Yunlin, Taiwan*

ABSTRACT:   In an era where purchasing a product no longer necessarily means a consumer's actual need for the product, there is a growing need for businesses to distinguish their brand from a plethora of others, create market advantages and segments, and boost the value of one's own brand in the consumers' mind. In order to achieve this goal, "design" is the main approach for creating brand value. As the global economic recession has intensified brand competition, striking a chord with consumers is necessary in order to arouse the consumers' desire to spend money in a discouraging economic climate. Branding would be the key to success, as it allows consumers to develop a preference resulting from acknowledgement. A sense of happiness, which is a concept all nations attach great importance to when making policies in relation to the happiness of a nation's people; however, few existing brands are designed according to consumers' sense of happiness. Therefore, this study proposes to determine what type of brand mascot, as well as what psychological, and spiritual elements a brand mascot must possess, in order to bring a sense of happiness and acknowledgment to consumers.

*Innovation in Design, Communication and Engineering – Meen, Prior & Lam (Eds)*
*© 2015 Taylor & Francis Group, London, ISBN 978-1-138-02752-7*

# US: A system to meet love by chance located in metro-bus transfer service

Zi Zhen
*School of Design and Arts, Beijing Institute of Technology, Beijing, China*
*Graduate School of Design, National Yunlin University of Science and Technology, Yunlin, Taiwan*

Teng-Wen Chang
*Department of Digital Media Design, National Yunlin University of Science and Technology, Yunlin, Taiwan*

Qi Luo
*School of Design and Arts, Beijing Institute of Technology, Beijing, China*

ABSTRACT: Transportation, especially modern metro-bus transportation process is a busy and lonely process for the passengers as well as the urban life. This research explores an interesting activity that might provide different aspects for such a routine process—encountering love. Using service design methods such as stakeholder and journey map, the daily activities of passengers are analyzed and classified into a set of routines that provide different routes towards chances to encounter love.

The main problem besides the shortest path during the transportation we believe are the mechanics and efficiency associated with the transportation itself that urban public transport often encounters, but how to solve this problem is not just an engineering issue, but an interaction design issue. What are the main characteristics of the users (man or woman)? And what are their desires is what we are aiming at. Through literature reviews, the love-matching issues are getting critical as the urbanization moves further in eastern society, especially modern China. One of possible meeting place is the transportation that is our contextual location. By studying encounters and matching social studies, we discover several interaction patterns of behaviors that are the base of our interaction design. A field site survey (we are doing observations and interview 11 people in Taipei and Kaohsiung) is conducted to verify the hypothesis and ensure the activities.

An implementation, called *us* is proposed and described in detail in a final paper. Service is built on an iOS platform. Interface and interactive usage scenarios will be displayed in the whole paper. These studies and designs for solving those problems include: "metro-bus transfer system", "urban men and women craving to encounter love" and the others aspects had a profound influence.

*Innovation in Design, Communication and Engineering – Meen, Prior & Lam (Eds)*
*© 2015 Taylor & Francis Group, London, ISBN 978-1-138-02752-7*

# Application of extensive QFD innovative procedure on railing design

F.G. Wu
*Department of Industrial Design, National Cheng Kung University, Tainan, Taiwan*

C.T. Wu
*Department of Industrial Design, National Kaohsiung Normal University, Kaohsiung City, Taiwan*

S.L. Luo
*Department of Industrial Design, National Cheng Kung University, Tainan, Taiwan*

ABSTRACT:   This research focuses on the application of extensive QFD innovative procedure for safe railing design. Railings are often installed in building windows or balconies for security, child safety, accident prevention, etc. but when a fire accident occurs, a railing will seriously affect a safe escape. Without an emergency exit, people will become "caged men" and the timing of escape and rescue will be affected. Therefore, the importance of emergency exits design of railings cannot be over emphasized. In this study, the extensive QFD developed by Wu has been adopted for an innovative design procedure. The major procedure of QFD is to identify the customers' needs for a product and then convert into appropriate technical characteristics. An innovative safety railing has been designed in this paper. The railings can be quickly disassembled into a rappelling escape ladder. Besides, the railings can also be extended as the platform waiting for rescue to avoid being hurt by smoke or fire.

*Internet technology*

*Innovation in Design, Communication and Engineering – Meen, Prior & Lam (Eds)*
© 2015 Taylor & Francis Group, London, ISBN 978-1-138-02752-7

# A load balanced and fault tolerant cluster management system from the aspect of house geomancy evaluation

Yu-Hsin Cheng
*Department of Information Networking and System Administration, Ling Tung University, Taichung, Taiwan*

Ming-Tsung Kao
*Department of Computer Science and Engineering, National Chung Hsing University, Taichung, Taiwan*

Yu-Sheng Chen
*National Chip Implementation Center, National Applied Research Laboratories, Taiwan*

Shang-Juh Kao
*Department of Computer Science and Engineering, National Chung Hsing University, Taichung, Taiwan*

ABSTRACT: In this study, a server cluster management system is established in a Kernel-based Virtual Machine (KVM) environment, using a load balancing mechanism provided by a mod_proxy of an Apache module and a fault tolerance mechanism through network card bonding technology. We use a Lu-Ban home Feng-Shui quantization and assessment service to verify the system's feasibility. The empirical implementation system consists of two servers, each equipped with three KVM virtual machines to accept requests from the single-entry front-end server, which is responsible for service allocation as well as source monitoring. Experimental results show that as the number of requests varies, this system can evenly distribute them among virtual machines in a cloud computing environment.

*Management science*

*Innovation in Design, Communication and Engineering – Meen, Prior & Lam (Eds)*
*© 2015 Taylor & Francis Group, London, ISBN 978-1-138-02752-7*

# A system dynamics investigation for the development and management of Chinese orchestra in elementary schools

Tian-Syung Lan, Pin-Chang Chen & Ying-Yan Lin
*Department of Information Management, Yu Da University of Science and Technology, Maioli County, Taiwan, R.O.C.*

Jai-Houng Leu
*General Education Center, Yu Da University of Science and Technology, Maioli County, Taiwan, R.O.C.*

ABSTRACT:   Chinese orchestra were spread from Mainland China to Taiwan after 1949. Under the active promotion of the government, Chinese orchestra have been widely accepted by various social classes and developed popularly. Because students participating in Chinese orchestra clubs have to purchase musical instruments and spend spare time on practice, this study intends to use system dynamics to investigate the influence of the price of musical instruments and afterschool practice time on the operation of Chinese orchestra clubs. The research results showed that, the price of musical instruments affects the willingness to participate in Chinese orchestra clubs. When the price of a musical instrument is low, the parents tend to have less burden and the students have higher willingness to participate in Chinese orchestra clubs. Moreover, the length of practice time also has a significant effect on learning effectiveness. Longer practice time leads to higher learning effectiveness.

# The impact of green accounting on product design and development

Jui-Che Tu & Hsieh-Shan Huang
*Graduate School of Design, Yunlin University of Science and Technology, Yunlin, Taiwan*

Chuan-Ying Hsu
*Department of Business Management, Dayeh University, Changhua, Taiwan*

ABSTRACT:   Since 2000, the Ministry of the Environment of Japan has implemented the guidance of environmental accounting. The Department of Sustainable Development of the UN (2001) and the International Federation of Accountants (2005) published the International Guidance of Environmental Management Accounting. In 2008, the Environmental Protection Administration also established the Industrial Environmental Accounting Guidance. Green accounting has become the mainstream of the world. Different countries require the exposure of environmental improvement information from industries. It is not only the issue of accounting, but also the topic related to product design and development. In order to accomplish the goals of "pollution prevention" and "design for environment", future implementation of laws and regulations will have a great impact on product design. Response of product design is the issue concerned by this study. By meta-research and literature review, this study probes into the effect of green accounting on product design. According to findings, design should be based on environmental protection, corporate social responsibility should be expanded, and product manufacturing must avoid pollution. External cost of production is internalized and re-design is conducted to improve product manufacturing and packaging. Balance is required for resources invested and products. The evaluation of life cycle of products is normalized. The development of environmentally friendly products is the trend to benefit the environment and increase competitive advantages. Green accounting system can provide figures as the reference for rewards or tax cutting.

*Materials for sustainable society*

# A study on a refurbishment model for existing North Paiwan slabstone of slabstone houses

Li-Mei Huang
*Department of Architecture, National Cheng Kung University, Tainan, Taiwan*
*Department of Interior Design, Tainan University of Technology, Tainan, Taiwan*

Chun-Ta Tzeng
*Department of Architecture, National Cheng Kung University, Tainan, Taiwan*

ABSTRACT: North Paiwanese slabstone houses are the oldest and lasting preserved aboriginal living spaces in Taiwan. After the retrocession, the Dailai, fawan, Majia villages possessed many traditional slabstone houses, which still maintain the original slabstone structures, and have the most renovated and refurbished slabstone houses compared to other North Paiwan villages. This study is based on the use of slabstone materials for sustainable society, researches the existing unique architectural style, expects to create a refurbishment model that can continually use the original slabstone bearing walls support the main structural space and explores a model that possibly combines "reused" and compatible materials with the local style and the traditional values of the existing slabstone houses. Through cluster analysis, this paper will analyze different types of reconstruction and refurbishments. The main factor is the use of local mountain materials and craftsman techniques instead of using materials from city and early self-made cooperation.

*Innovation in Design, Communication and Engineering – Meen, Prior & Lam (Eds)*
*© 2015 Taylor & Francis Group, London, ISBN 978-1-138-02752-7*

# Material innovation and application of vertical garden system in Taiwan

Li-Hsun Peng
*Department of Creative Design, National Yunlin University of Science and Technology, Yunlin, Taiwan*

Chia-Hsin Hsueh
*Graduate School of Design, National Yunlin University of Science and Technology, Yunlin, Taiwan*

ABSTRACT:   Nowadays, the vertical garden has become the architectural industry's best way of promoting innovative green material in Taiwan. Except for its visual effects which can attract the audiences, its originality of combining the green image of the architecture with the temporal traces of the traditional architecture brings out unique architectural renovation as well as the green impression. This study employs green design concept and materials science to show the principles of social, ecological and aesthetic sustainability of the vertical garden system in Taiwan. We realize that the reason we take Taiwan's vertical garden design as the main analytic subject is because of its uniqueness and necessity that represents one of the interpretations of using innovative green material in cultural and creative transition.

By using Phenomenography and case study as our main methodologies, this research will follow the use of green cluster as our research theories. It hopes to attain the following three goals: 1) analyzing green design theories in terms of material application on vertical garden system in Taiwan, 2) learning to recognize and value the environmental criteria and guidelines for green materials, manufacturing, and quality through an analysis of papers and cases, and 3) suggesting how green cluster might link to educational aspects and contribute to cultural and creative industries in Taiwan. We believe that this present study, which draws on phenomenographic research methodology, will help us maintain a positive attitude in developing a broader perspective on material application in connection with the creative vertical garden design in Taiwan.

*Mechanical & automation engineering*

*Innovation in Design, Communication and Engineering – Meen, Prior & Lam (Eds)*
*© 2015 Taylor & Francis Group, London, ISBN 978-1-138-02752-7*

# Analysis and experiment of performance of a fan-sink assembly

Shueei-Muh Lin & Wen-Rong Wang
*Mechanical Engineering Department, Kun Shan University, Tainan, Taiwan, R.O.C.*

Ming-Hong Lin & Sen-Yung Lee
*Mechanical Engineering Department, National Cheng Kung University, Tainan, Taiwan, R.O.C.*

Min-Jun Teng
*Guangzhou College of South China University of Technology, Guangzhou, China*

Shu-Jhang Lee
*Mechanical Engineering Department, Kun Shan University, Tainan, Taiwan, R.O.C.*

ABSTRACT: In this study, a cooling system composed of a double-entry centrifugal fan and a longitudinal-plate heat sink is investigated. The fan-sink assembly is popular for cooling the device of a notebook and some electronic systems. Due to the complexity of their coupling, the performances of a fan and a heat sink are investigated separately in the literature. Thus the overall cooling mechanism and the best cooling efficiency of the system have not been revealed. However, the coupled cooling system composed of a fan and a heat sink is investigated together. For the accuracy of CFD analysis, the overall flow field is investigated by implementing the κ–ε turbulent model. The optimum coefficients of the turbulent model are determined by comparing the numerical results and the experimental ones. Moreover, the influence of parameters of fin on the performance of the cooling system is investigated.

*Keywords:* centrifugal fan; fan-sink assembly; CFD; experiment; κ–ε turbulent model

*Innovation in Design, Communication and Engineering – Meen, Prior & Lam (Eds)*
*© 2015 Taylor & Francis Group, London, ISBN 978-1-138-02752-7*

# Enhancement of selective siphon control method for deadlock prevention in FMSs

Yen-Liang Pan
*Department of Avionic Engineering, Airforce Academy, Koahsiung, Taiwan*

Cheng-Fu Yang
*Department of Chemical and Materials Engineering, National University of Kaohsiung, Kaohsiung, Taiwan*

ABSTRACT:   One novel control policy named selective siphon control policy is proposed to solve dead-lock problems of flexible manufacturing systems. The new policy not only solves the deadlock problem successfully but obtains maximally permissive controllers. According to our awareness, the policy is the first one to achieve the goal of obtaining maximally permissive controllers for all S3PR models in the existing literature. However, one main problem is still needed to be addressed and solved in their algorithm. The problem is that the proposed policy cannot check the exact number of maximal permissive states of a deadlock net in advance. After all iterating steps, the final maximal permissive can then be known. Additionally, all legal markings are still to be checked again and again until all critical markings vanished. In this paper, one computationally improved methodology is proposed to solve the two problems. According to the experimental results, the computational efficiency can be enhanced based on the proposed methodology in this paper.

*Innovation in Design, Communication and Engineering – Meen, Prior & Lam (Eds)*
*© 2015 Taylor & Francis Group, London, ISBN 978-1-138-02752-7*

# PSO optimization Fuzzy-PID controllers applied to a Tri-rotor UAV

Shou-Tao Peng & Huu-Khoa Tran
*Mechanical Engineering Department, Southern Taiwan University of Science and Technology, Tainan City, Taiwan*

Juing-Shian Chiou
*Department of Electrical Engineering, Southern Taiwan University of Science and Technology, Tainan City, Taiwan*

ABSTRACT: Based on some sort of simplified fuzzy reasoning methods and PID parameters, many Fuzzy-PID controller schemes are applied to control the complicated system, recently. In this article a novel PSO algorithm, as an improved variant of stochastic optimization strategy PSO, which assigns optimization for Fuzzy-PID control gains, is introduced. The benefit integration design of the PSO algorithm structure generates and updates the new parameters for the Fuzzy-PID control schemes. The proposed controller has demonstrated a better performance to nonlinear Tri-rotor UAV control models.

*Keywords*: Fuzzy Logic Controllers (FLC); Proportional-Integral-Derivative (PID); Particle Swarm Optimization (PSO); Integral of Absolute Error (IAE); Tri-rotor UAV

*Innovation in Design, Communication and Engineering – Meen, Prior & Lam (Eds)*
*© 2015 Taylor & Francis Group, London, ISBN 978-1-138-02752-7*

# Dynamic simulation and application of virtual reality using dual axis device to remote control

Ko-Chun Chen
*Department of Applied Geoinformatics, Chia Nan University of Pharmacy and Science, Tainan, Taiwan*

ABSTRACT:   This paper's purpose of the virtual reality system are the environment and space limitations with convenience and safety features. Because the scenes actually have a high plasticity, they are widely used in the virtual reality system, such as entertainment, simulation and training, and so on. This research focuses on love rollercoaster games available to the operators of the game so users can feel the joy of roller coasters on the interior and higher irritation. The purpose of this study was the development of a double-axis dynamic simulator and next-generation platform using low cost control board and reversible motor. This control using the Bluetooth wireless signals can be exempt from the intricacies of the wire as well as allowing for platform devices wireless remote control and a number of weapons systems display purposes. Types of dynamic feedback models were first used in a virtual reality system, therefore, the environment and location needs to be adjusted in order to meet the operator.

*Media design & technologists*

Innovation in Design, Communication and Engineering – Meen, Prior & Lam (Eds)
© 2015 Taylor & Francis Group, London, ISBN 978-1-138-02752-7

# Analysis of the characteristics of digital image based on color

Yu-Bo Sa
*Xiamen Academy of Arts and Design, Fuzhou University, Xiamen, Fujian, P.R. China*

ABSTRACT: This paper summarizes the application of popular digital image mode according to the digital image color computer language description. It is based on extensive analysis, research and application of digital image color mode. It proposes specific application directions according to the characteristics of the digital image color mode with reference to international and industry standards of color management. Finally, it concludes the digital color image reproduction of reality is infinite. With the advent of digital technology, digital imagery is set to replace traditional imagery.

*Innovation in Design, Communication and Engineering – Meen, Prior & Lam (Eds)*
*© 2015 Taylor & Francis Group, London, ISBN 978-1-138-02752-7*

# Editing analysis on emotional expression of film

Xiao-Jing Yu
*Xiamen Academy of Arts and Design, Fuzhou University, Xiamen, Fujian, P.R. China*

ABSTRACT: Editing is the art throughout the creation of films, closely related to the feelings of the audience. Appropriate film editing can enhance the emotional expressions, and strengthen the feelings of the audience. Based on the theory of Montage, analyzing a number of classic clip points, this paper classifies fragments of different emotions and conducts the clip-point data processing and comparison. Identifying the relationship between the frequency of the clip points and the emotional expressions in films, this paper provides the data reference for film-video post-editing, and improves the accuracy of emotional expression editing, which can help to improve the emotional content in films.

*Innovation in Design, Communication and Engineering – Meen, Prior & Lam (Eds)*
*© 2015 Taylor & Francis Group, London, ISBN 978-1-138-02752-7*

# Changer—a project of contemporary image and body based on theory mirror stage

Saiau-Yue Tsau, Ko-Chiu Wu & Yu-Chun Huang
*National Taipei University of Technology, Taipei City, Taiwan, R.O.C.*

ABSTRACT:    Under the impact of globalization, the young generation has more choices to live their life; we often say farewell to friends who want to go abroad and pursue their so-called future to "Perfect my-self". According to Lacan's Mirror Stage theory (1936), we found that people consider their own image in the mirror and the image of other people as they were in their childhood; it's the same as when we grow up; how you will identify who you are is often dependent on where you have been and which friends you keep, based on the observations of the existing social conditions. This article proposes to look at a social problem which is "Modern people often lose themselves". Therefore, exploring how to deal with the psychological level of modern people as a concept of Changer. On the other hand, the side of technology aspect of this paper is focused on how to make dancers who need to play different roles, by changing clothes is normally considered as a change character for the audience. Following the observation of other performances, most theatrical dancers don't have enough time to change clothes. However, the stage background transform more rapidly than a dancer can change. Therefore, this paper will create an interactive platform and present an interactive art performance—Changer.

*Nanomaterials application*

# Effect of low-heating-rate vacuum annealing on aluminum doped zinc oxide films

Shang-Chou Chang & Wei Syuan Syu
*Department of Electrical Engineering, Kun Shan University, Tainan, Taiwan*

ABSTRACT: The Aluminum doped Zinc Oxide (AZO) films were vacuum annealed at a 10°C per minute heating rate. The AZO films' crystalline structure, surface morphology, carrier concentration, mobility, electrical resistivity and average optical transmittance were all probed for the AZO films before and after vacuum annealing. The (002) preferential direction of the vacuum annealed AZO films in X-ray diffraction spectra shifts to a lower angle. The carrier concentration, mobility and average optical transmittance increases while the electrical resistivity decreases for the vacuum annealed AZO films. The AZO films deposited at 200°C after vacuum annealing can reach $1.2 \times 10^{-3}$ ohm-cm in electrical resistivity and 92% in average optical transmittance.

*Keywords*: low-heating-rate; vacuum annealing; AZO

*Innovation in Design, Communication and Engineering – Meen, Prior & Lam (Eds)*
*© 2015 Taylor & Francis Group, London, ISBN 978-1-138-02752-7*

# Mg doping effect on the microstructural and optical properties of ZnO nanocrystalline films

S.L. Young, M.C. Kao, H.Z. Chen & N.F. Shih
*Department of Electronic Engineering, Hsiuping University of Science and Technology, Taichung, Taiwan*

C.Y. Kung & C.H. Chen
*Department of Electrical Engineering, National Chung Hsing University, Taichung, Taiwan*

ABSTRACT: Transparent $Zn_{1-x}Mg_xO$ (x = 0.01, 0.03 and 0.05) nanocrystalline films were prepared by sol-gel method followed by thermal annealing treatment of 700°C. Mg doping effect on the microstructural and optical properties of the $Zn_{1-x}Mg_xO$ films were investigated. From SEM images of all films, mean sizes of uniform spherical grains decreased progressively. Pure wurtzite structure was obtained from the results of XRD. Grain sizes increased from 34.7 nm for x = 0.01, 37.9 nm for x = 0.03 to 42.1 nm for x = 0.05 deduced from the XRD patterns. The Photoluminescence spectra of the films show a strong ultraviolet emission and a weak visible light emission peak. The enhancement of ultraviolet emission and reduction of visible emission is observed due to the increase of Mg doping concentration and the corresponding decrease of oxygen vacancy defects.

*Innovation in Design, Communication and Engineering – Meen, Prior & Lam (Eds)*
© 2015 Taylor & Francis Group, London, ISBN 978-1-138-02752-7

# Discussion for fuzziness on design of Chinese traditional farm tools

Jian Zhang
*Xiamen Academy of Arts and Design, Fuzhou University, Xiamen, Fujian, P.R. China*

Jing-Jing Huang
*Xiamen University of Technology, Xiamen, Fujian, P.R. China*

ABSTRACT: The design of Chinese traditional farm tools is the outward expression of traditional culture, and the study on design features of the traditional farm tools is the important means of the study on the traditional technology and civilization in China. At first, this article puts forward the "accuracy" and "fuzziness" concepts on the design of the traditional farm tools; and then the article expounds the reasons why the "fuzziness" will be generated through the analysis of the design and the material selection, as well as the production and forming processes of the traditional farm tools; as well this article discusses the specific performance in the five inspects, including structural components, structural proportion, the mechanical character, users and applications of traditional farm implements; furthermore, through the analysis for the design's "fuzziness" feature of the traditional farm tools, this article studies the cultural characteristics of traditional design in order to inspire the design field and promote the establishment of the design system in our country.

# Design of medical infanette based on service design idea

Jing-Jing Huang
*Xiamen University of Technology, Xiamen, Fujian, P.R. China*

Jian Zhang
*Xiamen Academy of Arts and Design, Fuzhou University, Xiamen, Fujian, P.R. China*

ABSTRACT:   Under the trend of designs transforming from "material" products to "immaterial" services, setting medical infanette as the object, this paper broke away from the traditional design approach of developing products from creativenesses and formed the "Product-Service" system combining medical infanette together with obstetric medical services based on the service design concept. Through the behavior and interaction researches among the three parties, namely, the newborn, the parturient and her families and the medical personnel, integrated the commonnesses and individuality demands of the three parties, put forward obstetric medical services model design based on medical infanette which provides references for forming a complete product concept, striving for creating healthy, scrupulous obstetric medical service experiences under the premise of meeting the demand of the three parties. It is one of the trends of medical product design in the future.

*Innovation in Design, Communication and Engineering – Meen, Prior & Lam (Eds)*
© 2015 Taylor & Francis Group, London, ISBN 978-1-138-02752-7

# Research on digital technology promoting the development of China's sculpture industry

You-long Pan
*Xiamen Academy of Arts and Design, Fuzhou University, Xiamen, Fujian, P.R. China*

ABSTRACT:    As the digital acquisition technology and digital molding technology have been gradually applied in China's sculpture industry, they can not only bring a technical reform to increase the productivity and lower the costs of both traditional handicraft sculpture industry and urban sculpture industry, but also create a new opportunity for the industrial upgrade of sculpture industry. This article will discuss how to utilize digital acquisition technology and digital molding technology to enhance the productivity of sculpture industry to a maximum extent.

# Rethinking atmosphere on a new trend of library spatial design under changing users' activities: Using the new library space of Yangtze University as case study

Jun He & Kuo-Hsun Wen
*Xiamen Academy of Arts and Design, Fuzhou University, Xiamen, Fujian, P.R. China*

ABSTRACT: As continuing development of information communication technology, libraries are experiencing a transformation from a book-centered to a user-centered place that leads to users' activities becoming more diversified. Consequently, the library interior space changes from simple to diverse; from close to open. It emphasizes more the possibilities of space adaptability, flexibility and replacement. Under the conception of "user-centered approach", the library design has to reveal affinity, exchange and atmosphere characteristics. This article employs the new library interior space of Yangtze University as case study to further interpret spatial design under the concern of changing activities from users.

*Innovation in Design, Communication and Engineering – Meen, Prior & Lam (Eds)*
© 2015 Taylor & Francis Group, London, ISBN 978-1-138-02752-7

# Library collection structure improvement based on statistic analysis of borrowing—taking Fuzhou University Academy of Arts & Craft Library as an example

Limei Zeng
*Xiamen Academy of Arts and Design, Fuzhou University, Xiamen, Fujian, P.R. China*

ABSTRACT: This article aims, taking Fuzhou University Academy of Arts & Craft Library as an example, through statistical analysis of borrowing in each major over the recent three years (2011–2013), at understanding the needs of the reader and analyzing reading psychology and psychological dispositions, for a better service for optimization of construction and collection, and college research. University Library is the information center of school's documents, an academic institution for teaching and scientific research. In the daily work of the library, serving readers are a priority, how to improve the utilization rate and satisfaction of the library is an important index of the library construction. By statistically analyzing the borrowing state and readers requirements in each major, adjusting the collection and rationally allocating library limited funds and resources, the borrowing status in Animation, Visual Communication, Environment Art and Industrial Craft is analyzed and the way to optimize the library collection structure by making use of "ILAS management system" of Fuzhou University Academy of Arts & Craft Library is discussed.

*Innovation in Design, Communication and Engineering – Meen, Prior & Lam (Eds)*
*© 2015 Taylor & Francis Group, London, ISBN 978-1-138-02752-7*

# Memory and electrical properties of (100)-oriented aluminum nitride thin films prepared by radio frequency magnetron sputtering

Maw-Shung Lee
*Department of Electronics Engineering, National Kaohsiung University of Applied Sciences, Sanmin District, Kaohsiung City, Taiwan, R.O.C.*

Sean Wu
*Department of Electronics Engineering and Computer Science, Tung Fang Design Institute, Hunei District, Kaohsiung City, Taiwan, R.O.C.*

Shih-Bin Jhong
*Department of Electronics Engineering, National Kaohsiung University of Applied Sciences, Sanmin District, Kaohsiung City, Taiwan, R.O.C.*

Kai-Huang Chen
*Department of Electronics Engineering and Computer Science, Tung Fang Design Institute, Hunei District, Kaohsiung City, Taiwan, R.O.C.*

Kuan-Ting Liu
*Department of Electronics Engineering, Cheng-Shiu University, Niaosong District, Kaohsiung City, Taiwan, R.O.C.*

ABSTRACT: In this study, the (100)-oriented Aluminum Nitride (AlN) thin films were well deposited onto p-type Si substrate by Radio Frequency (RF) magnetron sputtering method. The optimal deposition parameters were the RF power of 350 W, chamber pressure of 9 mTorr, and nitrogen concentration of 50%. Regarding the physical properties, the micro-structure of as-deposited (002) and (100)-oriented AlN thin films were obtained and compared by XRD patterns and TEM images. For electrical properties analysis, we found that the memory windows of (100)-oriented AlN thin films are better than those of (002)-oriented thin films. Besides, the interface and interaction between the silicon and (100)-oriented AlN thin films was a seriously important problem. Finally, the current transport models of the as-deposited and annealed (100)-oriented AlN thin films were also discussed. From the results, we suggested and investigated the large memory window of the annealed (100)-oriented AlN thin films, induced by many dipoles and large electric field, be applied.

*Innovation in Design, Communication and Engineering – Meen, Prior & Lam (Eds)*
© *2015 Taylor & Francis Group, London, ISBN 978-1-138-02752-7*

# The study of evaluation of family-group visitors of interactive exhibit patterns in science museums

Chen-Jung Yen
*Department of Visual Communication Design, Ling Tung University, Taichung, Taiwan*

ABSTRACT: This paper evaluates family-group visitors' attitudes toward interactive exhibits at the National Museum of Natural Science in Taiwan to examine their behavior processes, reactions, level of learning. This study hopes to provide museum visitors with a dimension of consideration on research and a reference frame for museum exhibit planning.

The research findings indicate that (1) most family-group visitors' visit the Science Center for entertainment, followed by education and learning purposes; (2) an increased length of visit influenced the frequency and content of parent-child interaction, although more static exhibits were unable to provide children with sufficient auditory or visual stimuli; (3) family visitors who stayed for shorter periods had better learning experiences than those who stayed longer; (4) a greater number of visits produced a lower level of satisfaction; and (5) visitors from Central Taiwan expressed greater satisfaction than visitors from other regions.

*Innovation in Design, Communication and Engineering – Meen, Prior & Lam (Eds)*
*© 2015 Taylor & Francis Group, London, ISBN 978-1-138-02752-7*

# Children's cognition of dangerous objects which cause fire

Fang-Suey Lin, Chun-Pei Hsieh & Ching-Yi Lin
*Graduate School of Design, National Yunlin University of Science and Technology, Yunlin, Taiwan*

ABSTRACT: Accidents are the main cause of death of children and teenagers. Due to incomplete cognitive development, children tend to be the victims of various kinds of accidents. In the families, an accident which causes serious harm and loss is fire. How can we guarantee children's domestic safety and avoid accidents? This study focuses on children's view of dangerous objects of fire, environmental cognition and unsafe behavior; the study uses 5–7 year-old kindergarten children as its subjects. Since children do not have a strong reading capacity, the researcher, in order to understand the factors of children's cognitive difference in this study, adopts pictures and images of dangerous objects as samples to conduct a survey on children's cognition of dangerous objects, which cause fire and dangerous behavior.. According to the findings: (1) children have a lower cognition of images of electric appliances which cause fire; (2) they tend to distinguish the images of inflammables and high-level cognition; (3) most of children have a primary cognition of the idea of not to play with fire; (4) children can associate dangerous objects with fire.

*Keywords*: children; fire; cognition; Domino Theory

# The influence of visual programming learning on logical thinking and reasoning ability for the sixth grade students in elementary school

Kai-Ming Yang
*Graduate Institute of Design, National Taipei University of Technology, Taipei, Taiwan*
*Department of Information Communication, University of Kang Ning, Tainan, Taiwan*

Chih-Heng Tsai & Ji-Yuan Wang
*Department of Information Communication, University of Kang Ning, Tainan, Taiwan*

ABSTRACT:    This study is mainly to investigate the influence of the logical thinking and reasoning ability after the sixth grade students of elementary school have learned visual programming. The findings of this study are the following: (1) students in the experimental group have more significant improvements on the logical thinking and reasoning ability; (2) the different learning styles of students in the experimental group have significant improvements on the ascensive progress of the logical thinking and reasoning ability. But there is no difference between the four types of terms of the students. (3) The learning attitude of students in the experimental group generally are high, like holding interest, willing to try and fix, patience for modifying the program, holding a positive learning attitude, and doing their works independently. (4) The students who have a better logical thinking and reasoning ability improved significantly on learning performance and logical thinking and reasoning ability test, after they learned how to use the visual programming design flow chart to help them understand the learning content.

*Innovation in Design, Communication and Engineering – Meen, Prior & Lam (Eds)*
*© 2015 Taylor & Francis Group, London, ISBN 978-1-138-02752-7*

# Quantitative research on hairdressing curriculum and workplace competencies of technical and vocational education

Hsiu-Tsu Chen
*Graduate School of Design, National Yunlin University of Science and Technology, Yunlin, Taiwan*
*Department of Fashion Design and Management, National Pingtung University of Science and Technology, Pingtung, Taiwan*

Kuo-Kuang Fan
*Graduate School of Design, National Yunlin University of Science and Technology, Yunlin, Taiwan*

Wei-Cheng Chu
*Department of Fashion Design, Shu-Te University, Kaohsiung, Taiwan*

Shu-Ping Chiu
*Department of Digital Media Arts and Design, Fuzhou University, China*

ABSTRACT: This study aims to investigate whether course contents of hairdressing related departments of technical and vocational education conform to workplace competencies required by the hairdressing industry, thus to know the importance of hairdressing curriculum planning in cultivating workplace competence in the hairdressing industry and provide a reference for education institutions to make adjustments and a plan of curriculum contents.

Using Delphi method and common decisions made by experts and scholars from industrial and academic circles, this study generalizes indicators for evaluating four-year college hairdressing courses and workplace competencies. By virtue of the consensus achieved through three rounds of expert questionnaires and the workplace competence indicators constructed, this study conducts a questionnaire survey among academic and industrial members, including 272 questionnaires sent out to hairdressing teachers and 382 questionnaires delivered to hairdressing practitioners. Findings show demands on hairdressing workplace competencies include five dimensions: work attitude, customer service, hairdressing knowledge and skill, personal development and marketing & operation management. Requirements of hairdressing curriculum planning are classified into four aspects: service ethics, hairdressing knowledge and skill, personal development, and marketing & operation management. Practitioners and scholars show differences in cognition of workplace competencies needed by the hairdressing industry. Compared with practitioners, scholars think there's a much stronger demand on "ability to plan and implement promotion activities", which is under the dimension of "marketing and operation management"; supervisors of the hairdressing industry have a significantly higher recognition of importance of "marketing and operation management" than designers. The present study demonstrates that demands on hairdressing workplace competencies are positively correlated with hairdressing course contents. These results indicate appropriateness of hairdressing curriculum planning influences students' future competitive capacity when entering the workplace.

*Innovation in Design, Communication and Engineering – Meen, Prior & Lam (Eds)*
© 2015 Taylor & Francis Group, London, ISBN 978-1-138-02752-7

# Exploring parents' experience and attitude toward their children in pain

Ching-Yi Lin
*Graduate School of Design, National Yunlin University of Science and Technology, Yunlin, Taiwan*

Fang-Suey Lin
*Visual Communication Design, National Yunlin University of Science and Technology, Yunlin, Taiwan*

Chun-Yi Lee
*Department of Pediatrics, Chang Bing Show Chwan Memorial Hospital, Changhua, Taiwan*

Chun-Pei Hsieh
*Graduate School of Design, National Yunlin University of Science and Technology, Yunlin, Taiwan*

ABSTRACT: Children's language, cognitive, and behavioral capabilities are still under development; children's experience of pain usually has to be conveyed by caregivers to the medical personnel, therefore, the communication process must further consider the parents' opinion, who may be the primary decision makers of the pain assessment. The rise of information technology not only changes our lifestyles, interpersonal communication patterns produce rapid changes due to the prevalence of Internet, and simultaneously has a significant impact in medical communication. This study explored parents' experience and attitude toward their children in pain and the use of mobile phone applications (apps) provide a new platform for delivering healthcare information. The semi-structured interviews are applied to this study and selected 17 parents who have children aged 3 to 7 years as research subjects. The aims of this study are to clarify the parentss dealing with an attitude of pain in children, children's experience in significant illness or injury and parents' views on children using technology products; the results of the interviews is conducted with grounded theory analysis. The present study finds that children generally experienced pain through sickness or accidents. When children experienced pain, and there are immediately observable external traumas or symptoms, then parents immediately seek medical attention. If there are no obvious symptoms, parents will first have a period of observation and reprocessing. In the care process, parents are in great need of nursing and medical care knowledge, while children need to be accompanied by others. Currently, most parents own smartphones and know how to use apps, so they can allow children to look at electronic touchscreen products and can accept the future application of these products in medical communications, but they also believe that there must be time restrictions. Therefore, it is necessary to understand the attitudes and needs of parents and children when they face pain, as a reference for the development of medical communications in the future.

*Keywords*: children; pain; parents; health communication

*Innovation in Design, Communication and Engineering – Meen, Prior & Lam (Eds)*
*© 2015 Taylor & Francis Group, London, ISBN 978-1-138-02752-7*

# Multiple interpretation of print technology artifact collection and exhibition

Chi-Shu Tseng
*Collections and Research Division, National Science and Technology Museum, Kaohsiung, Taiwan*

Yao-Ming Chu
*Department of Industrial Technology Education, National Kaohsiung Normal University, Kaohsiung, Taiwan*

ABSTRACT: This study uses the "Print Technology Artifact Exhibition" at the Natural Science and Technology Museum with 5 artifact collectors and exhibit developers as subjects using qualitative research and in-depth interviews along with related documents to understand the differing interpretation of perspective by collectors and exhibit developers towards technological artifacts in order to promote mutual understanding and communication. Data will be processed with the principles of thematic analysis using "whole—part—whole" process to view research data and categorize. The study will then construct a two-dimensional structure of interpretation according to relevance. The study has reached the following conclusions: 1. Collectors pay less attention on the facets of the object itself and structure when interpret print technology artifacts. 2. Exhibit developers interpret artifacts lack of the facets of industry history and social culture. 3. The interpretation of technological objects by collectors and exhibit developers intersect on the facets of technological development and function.

*Innovation in Design, Communication and Engineering – Meen, Prior & Lam (Eds)*
*© 2015 Taylor & Francis Group, London, ISBN 978-1-138-02752-7*

# The effect of learning chaos concept for high school students by attending IYPT

C.H. Yang, J.S. Lih, T.C. Chen, M.C. Lu & M.C. Ho
*Department of Physics, National Kaohsiung Normal University, Kaohsiung, Taiwan, R.O.C.*

C.M. Hung, Y.S. Liu & C.J. Liu
*Graduate Institute of Science Education and Environmental Education, National Kaohsiung Normal University, Kaohsiung, Taiwan, R.O.C.*

ABSTRACT: This study aims to investigate the feasibility of teaching nonlinear physics courses in high school by attending International Young Physicists' Tournament "IYPT", a competition with physics debates on building a Chua's Circuit that displays chaos behaviors. The participants comprise high school teachers and five high school students studying in a science program and undergoing three stages of exploring and learning. Meanwhile, the students learnt to understand the properties and behaviors of chaos by simulation and building a nonlinear Chua's Circuit with basic electronic components. Thus, they were able to present their studies, question others about their studies, and summarize the key points during the debate competition. Results show that chaos teaching in high schools is feasible and improving the professional knowledge of nonlinear physics for current high school teachers is necessary.

*Innovation in Design, Communication and Engineering – Meen, Prior & Lam (Eds)*
*© 2015 Taylor & Francis Group, London, ISBN 978-1-138-02752-7*

# A study on the application of big data analytics in design

Qing Lin
*Xiamen Academy of Arts and Design, Fuzhou University, Xiamen, Fujian, P.R. China*

ABSTRACT:  With the advent of the era of big data, design faces unprecedented changes and challenges. The era of big data has also created many more opportunities. Data is gradually becoming a major design resource, and big data is transforming how we think. The big data overlay analytics and visualization analytics are increasingly utilized in design.

*Innovation in Design, Communication and Engineering – Meen, Prior & Lam (Eds)*
*© 2015 Taylor & Francis Group, London, ISBN 978-1-138-02752-7*

# AI techniques and its application in game programming

Jun He

*Xiamen Academy of Arts and Design, Fuzhou University, Xiamen, Fujian, P.R. China*

ABSTRACT: In games, players are not always interested in giving non-player characters of human-level intellect. Usually, game producers write codes to control non-human creatures such as dragons, robots, or even rodents. Artificial Intelligence (AI) in game programming needs further study to solve fairly complex problems. However, if a Non-People Character (NPC) is used for attempting to have different personalities and unusual appearances with emotions, fear, and disposition, the effect of entertainment and the variety of game content is expected to be much stronger. At present, the artificial intelligence interest level in the game industry is relatively low owing to restrictions of existing level, hardware resources. But recently, game developers, producers and managers reckon artificial intelligence application as one of the important factors for game programming. This study analyses the behavioral course of NPC applied genetic Artificial Intelligence Method to maintain the players' interest and provide the way to enhance their interest. It demonstrates designed and realized game NPC behavioral course is used properly in the game map.

*Thin film & device process*

*Innovation in Design, Communication and Engineering – Meen, Prior & Lam (Eds)*
© *2015 Taylor & Francis Group, London, ISBN 978-1-138-02752-7*

# Properties of hydrogen plasma treated Gallium and Aluminum co-doped Zinc Oxide films

Shang-Chou Chang & Hung Jen Chen
*Department of Electrical Engineering, Kun Shan University, Tainan, Taiwan*

ABSTRACT: Gallium and Aluminum co-doped Zinc Oxide (GAZO) films were prepared by in-line sputtering at room temperature. After that, the GAZO films were treated with different treatment time of microwave hydrogen plasma. The treatment time of microwave hydrogen plasma varied from 1 to 10 minutes. The surface roughness of GAZO films increases with the treatment time and was observed from scanning electron microscope. The increase of surface roughness can be attributed to re-crystallization of GAZO films after microwave hydrogen plasma treatment. Microwave hydrogen plasma annealing can improve the electrical and optical properties of GAZO films. The $5.6 \times 10^{-4}\Omega$-cm in electrical resistivity and 93% in average optical transmittance with visible wavelength region can be obtained for GAZO films after microwave hydrogen plasma treatment of 5 minutes.

*Keywords*: plasma treatment; GAZO

*Innovation in Design, Communication and Engineering – Meen, Prior & Lam (Eds)*
*© 2015 Taylor & Francis Group, London, ISBN 978-1-138-02752-7*

# Growth characteristics and transparent properties of Al-Si-O films prepared by E-beam evaporation on PET substrate at room temperature

Hong-Hsin Huang
*Department of Electrical Engineering, Cheng Shiu University, Kaohsiung, Taiwan*

Cheng-Fu Yang
*Department of Chemical and Materials Engineering, National University of Kaohsiung, Kaohsiung, Taiwan*

ABSTRACT: The transparent and protective thin Al-Si-O films had been prepared by E-beam evaporation method on flexible PET substrate. The effects of Al/Si ratio and evaporation pressure on growth characteristics and transparent properties were investigated. The structure, surface morphology, composition, and transmittance were characterized by the X-Ray Diffraction (XRD), scanning electron microscopy, energy dispersive spectrometry, and UV–Visible spectroscopy. Results showed that amorphous thin films were obtained although a broaden peak was found in the XRD pattern of thin film prepared at low pressure. As pressure increasing, the morphology changed from island to granular growth, in addition, the fine granular size was found as the Al/Si ratio increased. High transmittance was found when thin films prepared at various pressures and with various Al/Si ratios. The optical band gap is similar about 3.095 eV, however, it increased with evaporation pressure as Al/Si = 1.

*Innovation in Design, Communication and Engineering – Meen, Prior & Lam (Eds)*
© *2015 Taylor & Francis Group, London, ISBN 978-1-138-02752-7*

# Optical characteristics and microstructures of large-field-angle P-polarizer deposited by E-beam system

Chih-Hao Zeng & Chien-Yue Chen

*Electronic and Optoelectronic Engineering in Microelectronic and Optoelectronic Engineering, National Yunlin University of Science and Technology, Yunlin, Taiwan*

Po-Kai Chiu, Chao-Te Lee, Donyau Chiang, Chien-Nan Hsaio & Fong-Zhi Chen

*Instrument Technology Research Center, National Applied Research Laboratories, Hsinchu, Taiwan*

ABSTRACT: We fabricated a multi-layer structure composed of $MgF_2$ and $HfO_2$ thin films deposited alternatively by an E-beam system on a fused silica. The multi-layer structure is designed for a kind of polarizers with large field angles, which are quite different from the common applications for a flat polarizer with normal incidence. The effects of the different ion-beam voltages of IAD deposition and the deposited parameters of $HfO_2$ on the optical characteristics and microstructures of the formed structure were investigated. The transmittances of an obtained structure measured in air at an angular field of 75° for a P-polarized and S-polarized light are greater than 90% and less than 3%, respectively, throughout the spectral region extending from 290 nm to 320 nm. The relative illuminance of the multi-layer sturucture is confirmed by the numerical simulation to be over 90% when the resultant structure is viewed at 75° tilted angle for a wavelength of 313 nm. The respective thicknesses of each layer and surface roughness of the multi-layer structure were measured by Scanning Electron Microscope (SEM) and Atomic Force Microscope (AFM).

*Innovation in Design, Communication and Engineering – Meen, Prior & Lam (Eds)*
*© 2015 Taylor & Francis Group, London, ISBN 978-1-138-02752-7*

# Author index